A NEW WATER FUTURE
Pacific Markets

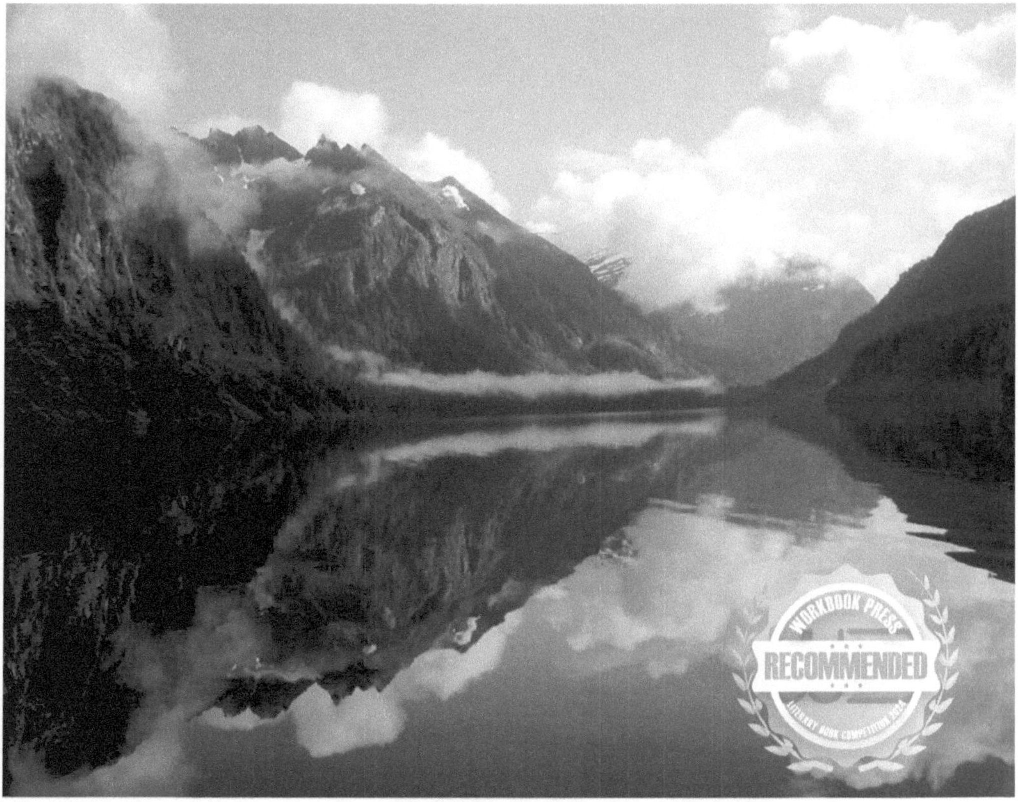

CALIFORNIA, MEXICO, SOUTH AMERICA, & CHINA

RIC DAVIDGE, MPA/PM   AUTHOR
FORMER ALASKA STATE DIRECTOR OF WATER

A NEW WATER FUTURE
Pacific Markets

California, Mexico, South America, & China

Copyright © 2024 by Ric Davidge

All rights reserved.
No part of this book may be reproduced in any form or by any electronic or mechanical means, including information storage and retrieval systems, without permission in writing from the publisher, except by reviewers, who may quote brief passages in a review.

This publication contains the opinions and ideas of its author.
It is intended to provide helpful and informative material on the subjects addressed in the publication.
The author and publisher specifically disclaim all responsibility for any liability, loss or risk, personal or otherwise, which is incurred as a consequence, directly or indirectly, of the use and application of any of the contents of this book.

WORKBOOK PRESS LLC
187 E Warm Springs Rd,
Suite B285, Las Vegas, NV 89119, USA

Website: https://workbookpress.com/
Hotline: 1-888-818-4856
Email: admin@workbookpress.com

Ordering Information:
Quantity sales. Special discounts are available on quantity purchases by corporations, associations, and others. For details, contact the publisher at the address above.

Library of Congress Control Number:

ISBN-13: 978-1-961845-12-1 (Paperback Version)
          978-1-961845-13-8 (Digital Version)

REV. DATE: 01/12/2023

Table of Contents

In Brief 1
EXECUTIVE SUMMARY 3
   The Facts 3
   The Fictions 4
1. INTRODUCTION – The Demand 7
2. BACKGROUND 17
   2.1   History of the bulk water business 17
   2.2   History of this endeavor in South East Alaska 21
   2.3   History of AQUEOUS International, Inc. 23
3. Rationale for this venture 33
4. TARGETED MARKETS 37
   4.1   China 38
   4.2   China's Market Outlook 48
5. Our Top Ten Market Targets in China 51
   5.1   Shanghai — 23.4 million people (first market target) 51
   5.2   Beijing — 18.8 million people 52
   5.3   Tianjin — 12.8 million people 53
   5.4   Shenzhen — 12.7 million people 54
   5.5   Guangzhou — 11.6 million people 55
   5.6   Chengdu — 10.2 million people 56
   5.7   Chongqing — 8.5 million people 57
   5.8   Dongguan — 8.3 million people 58
   5.9   Shenyang — 7.9 million people 59
   5.10 Wuhan — 7.9 million people 60
   5.11 A Brief Market History – this decade 61
   5.12 UPDATE: 63
   5.13 South Korea: 65
   5.14 Water Transport Bags (WTB) from S Korea to Chinees markets .
   68
   5.15 China Complications 69
   5.16 India 71
   5.17 Middle East 81
   5.18 California/Nevada et al 83
      5.18.1   The Jones Act is a federal law 83
   5.19 The Global Market Perspective 84
   5.20 Other Strategic Market Reports 86
   5.21 Bulk Water to the United States 86
      5.21.1   Water Market USA 87

5.22	Bottled Water: A Global Strategic Business Report	88
6.	Why Alaska *natural* Glacial Water now?	101
6.1	Why water from Alaska, and why twenty sources to start?	101
6.2	Our Southeastern Alaskan Sources	103
7.	PRODUCT DESCRIPTION	109
7.1	Water Quality	112
7.2	What we provide to markets	113
8.	BULK CONVEYANCE OPTIONS	117
8.1	Barge	117
8.2	General Bulk-Carrier	118
8.3	Tanker Ship	118
8.4	PREFERED Tanker Option	119
8.5	Water Transport Bags (WTB)	119
8.5.1	Loading	122
8.5.2	Panamax Class Tankers	122
8.5.3	Aframax Tankers	124
8.5.4	VLCC Class Tankers (not a VMaxx)	125
8.5.5	VMaxx VLCC class tanker (preferred)	126
8.5.6	The V Maxx:	126
8.6	ULCC Class Tankers	129
8.7	Tanker Availability	130
8.8	Distances for Conveyance from Southeast Alaska	135
8.9	Delivering "Drinkable Water"	137
8.10	Price Targets at Market/Point-of-Sale	138
9.	STRATEGY	141
9.1	Business Models (two)	141
9.2	Short Term (5 to 7 years)	143
9.3	Long Term 50-year Plan	147
10.	DELIVERY CONTRACTS	151
10.1	Delivery Contracts are essential.	153
11.	COMPETITION	157
11.1	Direct Competition	157
11.2	South Korea	159
11.3	Other sources of water for global markets	161
11.4	The Costs of Options	163
12	DEVELOPMENT PLAN	165
12.1	Key Elements of the Project – and anticipating questions	165
12.2	Loading/Off Loading limitations	166
12.3	Shipping – additional notes	167
12.4	Energy costs	168
12.5	Using Ballast tanks for freshwater	168

12.6	Other Important Cost Variables when Tankers are Purchased	169
12.7	Hyundai Heavy Industries – a VMaxx Proposal	170
12.8	Storage	170
12.9	Distribution	171
12.10	Development & Engineering Plan	171
12.11	Permits/Licenses	172
12.12	Loading and Off-loading permits	172
12.13	Shipping Licenses	172
12.14	Description of what needs to be done	173
12.15	It is Time (first three years)	174
12.16	Items to be done to achieve goals	179
12.17	How proven is the technology?	181
12.18	Options to each significant task	181
13	MARKETING	183
13.1	International Meetings	183
13.2	Publications	185
13.3	Presentation	186
14	BUSINESS DEVELOPMENT	189
15	MANAGEMENT	191
15.1	Management Team	191
15.2	Project Development Team	191
15.3	Contractors	192
16	FINANCIAL PLAN	195
16.1	Legal Structure	195
16.2	Capital Structure	195
16.3	Milestones and Objectives for multiple funding rounds	196
16.4	Budget	197
16.5	Initial Investors	198
16.6	Capital Investments	198
16.7	Intellectual Property Investments	198
16.8	Detailed cost and profit breakdown/Project Proformas	200
17	LEGAL	201
17.1	Water Rights and Contracts	201
17.2	Ability to Export	201
17.3	Environmental Issues	202
17.3.1	Local	202
17.3.2	International	202
18	RISKS	207
18.1	Federalism	207
18.2	Other Risks	208
18.3	Legal Problems	209
18.4	Adverse Media	209

TABLE of FIGURES

Figure 1:	Bottled Water in China Financial Illustration	1
Figure 2	Drinking Water: A Global Context	16
Figure 3:	Projected Water Scarcity	35
Figure 4:	Dead Infected Pigs litter Shanghai Beach.	39
Figure 5:	Pollution	42
Figure 6:	Pollution Clogged River.	43
Figure 7:	Initial Market Information.	45
Figure 8:	Chinese Pollution.	46
Figure 9:	Shanghai.	51
Figure 10:	Beijing	52
Figure 11:	Tianjin.	53
Figure 12:	Shenzhen.	54
Figure 13:	Guangzhou.	55
Figure 14:	Chengdu.	56
Figure 15:	Chongqing.	57
Figure 16:	Dongguan.	58
Figure 17:	Shenyang.	59
Figure 18:	Wuhan.	60
Figure 19:	"Know Your Market".	63
Figure 20:	Chinese Bottled Water.	64
Figure 21:	Chinese Bottled Water.	64
Figure 22:	Eastern Markets.	66
Figure 23:	Chinese Port.	67
Figure 24:	South Korean Shipyard.	68
Figure 25:	Bottled Water of India.	73
Figure 26:	Growth in Bottled Water Demand in India.	76
Figure 27:	Global Bottled Water Market 2013-2021.	88
Figure 28:	Global Bottled Water Market Split by Region (2017).	89
Figure 29:	Development Process Flow Chart	107
Figure 30:	Water Conversion Table.	111
Figure 31:	WTB Testing	121
Figure 32:	Panamax Class Tankers.	122
Figure 33:	Aframax Tanker.	124
Figure 34:	ULCC Class Tanker.	129
Figure 35:	Conveyance Distances for Southeast Alaska	135
Figure 36:	Aframax Class Tanker Costs	135
Figure 37:	Aframax Per unit Costs	136
Figure 38:	Two Business Approaches	141
Figure 39:	Key Numbers	149
Figure 40:	Delivery Contract Stipulations	151
Figure 41:	Delivery Contract Buyer Role	152
Figure 42:	Cost Variables when Tankers are Purchased	169
Figure 43:	What Needs to be Done	174
Figure 44:	Costs to be Identified	180
Figure 45:	Immediate Development Needs	188
Figure 46:	Budget- Employment Contractors	197
Figure 47:	Capital Investments	198

In Brief

In **China**, premium **bottled water** typically **costs** about 10 yuan per bottle, compared to the **price** of mass market brands, which range from 1 to 3 yuan. The gross profit margin for the premium brands is about 70%, compared to approximately 30% for mass market brands.

2020 Wholesale price of <u>filtered</u> water in China is:

$.12 per gallon

Price Targets for first market

China
Our cost at delivery $.07 per gallon
Startup price $0.9 per gallon China (increase once established in market)

Illustration
1 water source
1 VMaxx lifts 90,000,000 gallons per lift at source
1 lift per month
Company income at $.01 per gallon at source per month
  Yields $900,000 per month <u>net</u> profit
  Yields $10,800,000 per year net profit

5 water sources
3 VMaxx per source (15 vessels) allows weekly deliveries to China again lifting 90,000,000 gallons per vessel per week
Company income at $.01 per gallon w5 sources X15 vessels yields 90,000,000 x $.01x 15 vessels x 50 weeks = $22,500,000,000

Figure 1: Bottled Water in China Financial Illustration

This is just an illustration using a minimum income per gallon at $.01. Our projections are that we have a margin in the wholesale market of $.03 to $.10 per gallon based on latest numbers in the ten Chinese markets.

Debt reduction, even with a VMaxx costing $150M each and a site infrastructure of under $1M, we can be paid out in just a few years, reducing cost per gallon (with no debt) significantly increasing net income.

NOTE: This book was completed just before COVID hit. Then we had to stop as I faced 3 surgeries that took a total of 13 months due to recovery time between each. The good news is that I'm completely clear now of any cancer, but my knee surgery didn't work. I'm not doing that again.

I've shared this so you will understand that some of the dates here are old, but we are now back in the saddle and moving forward on source security.

The national and state elections could accelerate our work or show it down. We monitor this daily.

Ric Davidge, MPA/PM
Water Czar

EXECUTIVE SUMMARY

The Facts

- Global demand for clean *natural* water has and continues to accelerate and increase in value.
- As we predicted over 30 years ago, commodity trading of bulk water has begun
- China is now the largest consumer of bottled water in the world and growing.
    - Bottled water is an <u>indicator</u> of freshwater demand growth
    - Ten targeted markets in China offer significant immediate opportunities.
- Mexico is very thirsty, closer, and cheaper. So, we start there and grow.
- Actual sustainable source security outside of America <u>is not available</u>.
    - Not likely to change due to political, cultural, and religious attitudes.
- Alaska has tremendous sustainable pristine *natural* even glacial freshwater to meet demand.
    - The most legally secure pristine *natural* water sources in the world are in Alaska.
    - Drinking water from 10K to 30K year old glacial sources provides unique BRAND value.
- Aqueous uniquely has the strategic bulk water knowledge, business model and structure to deliver enormous quantities of pristine *natural* Alaskan glacial water to global markets cost effectively with solid margins in developed markets that are sustainable for generations.
- All water contamination in surface, ground, and saltwater is accelerating globally.
    - Pharmaceuticals, fecal coli, industrial wastes, agricultural waste, etc.
        - **the chemistry of water has changed in most of the world, even in America**
    - Growing personal awareness of the extraordinary and increasing levels and types of contamination, especially pharmaceuticals (estrogen is a serious concern).
    - Accelerating personal wealth by those most interested in healthy water.
    - The cost of cleaning freshwater for specific industries dependent on quality water increases and removes essential minerals for products.
    - Growing awareness that desalination is not the answer due to

- adverse health effects especially for an embryo and the young (all species).
  - Waste discharges from desal globally recently discovered to be 50% more than the industry reported and seriously threaten ocean species and greatly raise membrane cost
  - With new technology, the cost of bulk water conveyance is significantly down per unit.
  - The economics work.

The Fictions

- Desalination will solve the problem.
  - Growing public awareness of the health risks to infants and most living organism's dependent on the minerals in their drinking water – especially an embryo and the young.
  - The energy cost of desalinated/demineralized water is high and very unstable.
  - Many desal plants are now intaking their own waste discharge – raising their cost
  - Global waste discharge volume is greater than 50% per system of what was expected harming ocean life and creating vast dead zones in many oceans
- Wastewater recycling will solve the problem.
  - Recently discovered magnitude of pharmaceuticals and the formation of new unknown compounds in vegetables grown with recycled water – no technology to remove them that is anywhere close of economically viable – and all of them remove essential minerals. Science is uncertain what the consumption of these mean to humans and other mammals.

The time is now, the best, most *secure natural* glacial freshwater in the world is in Alaska, and Pacific markets are available and thirsty.

Rio Davidge, MPA

1. INTRODUCTION – The Demand

Freshwater Crisis

The global freshwater crisis in general, caused by continued growth in populations and their migrations, the growth in personal demand as economies mature, the growing and consistent demand by industry for more and higher quality freshwater (filtration costs are raising quickly), and the accelerating increase in all spectrums of water contamination especially pharmaceuticals like estrogen, continues to open serious significant business opportunities in global markets for competent capable future minded investors. But water investors must not only understand the opportunities in targeted markets, but they must also be capable of this level, scale, and duration of investment. They must understand the technological and systemic applications that work in three to five years generating <u>significant</u> and stable generational returns. They must understand the globally unique legal framework provided by Alaska for 'source security'. The international bulk *natural* freshwater demand is already greater than one hundred billion a year, in US dollars, and AQUEOUS International, Inc. is the only entity with the knowledge and capability to bring at least 90 Billion gallons a year of pristine *natural* Alaskan glacial freshwater to Pacific Asian markets starting in just three years. Yes, we are interested in other global markets and have other pristine natural water sources assessed in or close to those markets, but this is where we have chosen to start. Why? Because the numbers work, and we have been tracking them for a long time. Because we have source security and both state and national protection of this 'owner right'. These are the keys.

United Nations Report

Although the United Nations report on the pending global freshwater crisis targeted 2025 as the tipping point [when there are more people without adequate water then with it, regardless of quality], our review of that and other similar international reports, revealed that the tipping date was much closer to 2015. This is mainly because the initial U.N. report did not consider or even understand the rapid, almost exponential decline of coastal aquifers (where most people and industry choose to be) due to over exploitation of coastal groundwater. Nor the rapid mutation of pharmaceuticals in ground, surface, and ocean waters. Once an aquifer is contaminated, especially with sea water due to hydrologic pressure, it will always be contaminated and will have to be treated or desalted/demineralized for any human use. At present, all coastal aquifers in India, China, most island nations and even portions of southern California and the Gulf of Mexico states are contaminated or will be within the next five years. **This trend is not reversible.**

Contamination & Collapse

Beyond the contamination of coastal aquifers, many inland aquifers have collapsed or will collapse over the next decade because of the incompetent management of take. Once they collapse and the spaces within which water from the glacial age (10,000 years ago) are pressed out and thus slowly collapse, it is next to impossible to 'recharge' or 'reopen' these subsurface voids to allow water to again accumulate. The southwestern region of the United States has seen a surface subsidence of over twelve feet (yes, the surface of the earth has changed) in most areas and in some locations more than twenty feet. This is not a unique situation, but one that is globally occurring as we pump the earth dry of its groundwater. All trends in the depletion and/or contamination of freshwater sources are pointing in one direction – scarcity. Given the limited resupply options, not good.

Tipping Point

We passionately believe that the tipping point was 2015 not 2025, and here we are in 2021. AQUEOUS International, Inc. understands these trends and monitors them daily. As a result, we know the causes and thus anticipate market demand all over the world and thus the opportunities for a global bulk water market that will result in the global commodification of freshwater just like what happened to oil. We are ready.

Economic Sensitivities

One of the other factors not adequately understood by these government reports in the global freshwater arena is the lack of understanding that economies are far more sensitive to either the increased cost of freshwater or its depletion to a point that they are no longer economically viable and then collapse. This single economic reality is what also pushed the tipping point far closer than 2525.

Already in 2011, we saw economies "adjusting" to less water or more cost to clean the water they receive to meet even minimum standards. Government subsidies for both homes and business continue to mask the real cost and value of this water not allowing markets to even see this reality in time.

One of the underlying factors here is the refusal by political boards and commissions in states and cities to charge for 'actual cost' for drinkable freshwater. This behavior perverts any sense of cost reality in economic terms. Essentially all municipal and state water utilities operate with subsidies – some exceptionally large subsidies, so the 'payer' has no real sense of the true cost of their water or of its value to them or their economy. This is an old soap box for us, but it is starting to get some attention by elected and appointed officials who are beginning to be honest about cost and at least start talking about 'actual cost' let alone value in their rates.

The Right Place at the Right Time

The possibility of this venture being in just the right place at just the right time is very real, but we have been studying this for many years. Read the chapter on the history of AQUOEUS International, Inc. in this document to give you context. It will help you understand our persistence. Our timing in acquiring the twenty water sources we have assessed in southeast Alaska is timely. We have already secured some sources but believe we must have at least twenty sites to accommodate the Southeast Alaska weather, fishing seasons, minus tides (up to 20 feet) etc. that we must work around to access our sources when a tanker calls – and to be fully established in front of global demand that we are confident will accelerate with our success.

Five Year Ramp-Up

Further, we predict that it will take about five years for this venture to establish its global image, market and source and conveyance credibility, and the capabilities to capture hundreds of immediate and emerging markets around the world, yes starting in the Pacific, at our necessary scale. That puts this company exactly at the right place at the right time to take advantage of the 'tipping point' and other growing global market demands.

Because of our collective experience in the development of a global water market, we now better understand that one approach to a market's freshwater needs may likely not be responsive to the opportunities in every market situation given the wide range of variables for each potential project. Bulk water imports can work in specific markets, but new technologies promising lower cost desalination either on or offshore may better respond to other markets (at least until they really understand the health cost of demineralized water and the ecological cost of increasing (recently discovered to be 50% greater then what the industry reported - brine discharges to sea life and growing RO process costs), and new systemic approaches in some markets dramatically enabling conservation - still the cheapest means of creating "new water" may make sense – and there are or may be new technologies that can capture these markets. The problem is that in almost every case these new technologies 'manufacture' their water which is not even close to the health and perception value of pure *natural* water with all its *natural* minerals. Not in taste but in the actual adverse health impacts on human beings (during pregnancy and the first years of mental development) and essentially all other living things. The depravation of natural minerals for personal ingestion has finally underwriting the general sense that we need to drink good *natural* water – not desal or manufactured/filtered waters that take out exactly what we need.

Mineral Deficiencies
There are some who argue that our food provides us with sufficient minerals to replace what may be lost by desal. But we know better as our soils, not just in America but globally, are losing their mineral content via several means including agra supplements that contaminate not just the soil but also the water, and the loss of topsoil in areas long farmed. There is much more here to discuss.

Localized Market Challenges
Each market presents specific and multi-variable challenges and opportunities. Those who best understand this and can meet the specific market's needs with secure sources at competitive prices, are those who will capture these markets, and keep them for decades (delivery contracts are generally 30 years, some 99 years), and watch profits grow. This is exactly why we are in search of 'market partners' not just buyers. The development of a bulk water market

in solid long-established high-end bottled water regions is the other side of our formula. The relationship between AQUEOUS International, Inc. and market partners are a key to our business plan. Each will be different, but our preference is for our market partners to provide their own conveyance, to deal with all their own political issues outside of our participation, to expand their markets to our advantage, and to work with us on marketing and brand not only as a base for personal drinks but also industry and maybe even municipal mix at some level.

Volume Business
Water is a volume business. That is how you make money, not one case at a time. Therefore, we have continued to explore and will develop new market partners in other than packaged drink markets for our premium *'fine'* water with industry and possibly municipal mix to lower their own contamination without more investment and the ever-present O&M cost of cleaning it – while also removing the natural minerals that need to be in healthy water.

Vivendi Water
The only company to even come close to succeeding in this global water market was **Vivendi Water** of France. Although they continue to be a big player, their focus, due mainly to generational change in ownership/management, of late has changed into entertainment and other investment areas, some painfully not successful. They also tend, as do all the others, to focus on pushing specific water filter technologies, that they own and sell, and are often unable to look beyond these applications and their adverse health aspects because of this interest to a broader market opportunity embracing other approaches. They have been in business for over 150 years and are an important likely competitor in the development and approach of AQUEOUS. But we have the advantage of <u>secure</u> pristine *natural* Alaskan glacial freshwater. They still take water from sources that have been peed in for over 2,000 years. Their new technologies that may be game changers in desalination markets are now facing growing public understanding of the provable human risk of consumption of demineralized water – especially the young. We start and stay with <u>*natural*</u> freshwater from glaciers of 10,000 to 32,000 years of age. This changes our attraction in markets as we push a different top-of-mind 'need' for *natural* pristine Alaskan glacial water and thus huge market advantages in all applications.

Business Models

This **b**usiness plan explores business models or approaches to the markets we have and continue to study with our partner Euromonitor and others. Population growth, migration especially rural to urban, new, or growing industry and their accelerating water cleaning costs, and general economic growth have pushed much of Asia into our arms – they are ready for us. We are unique in our source security, extraordinary *natural* water quality, and thus huge market potential due to ever more enlightened consumer demand stimulated by our brand marketing. The bulk export of almost 100 Billion gallons a year of Alaskan *natural* glacial water from southeastern Alaska, as our principle and importantly most secure sources, places an **extraordinarily high quality** and **unique product** into the global water market. It is not meant for public water systems, although it could be used as "mix" to increase municipal water quality in some unique markets with significant contamination issues.

Target Audiences

This water product is specifically targeted for markets that understand the unique opportunity of high quality or "fine" **bottled water or beverages**, still the highest per unit ROI in a variety of retail packages to include one gallon or smaller "personal" packaging. Unlike the United States, in most overseas markets, **the use of bottled water is the preferred or only means of getting safe drinking water** due to local public water quality concerns or availability. This also enables the price per unit at retail to absorb the cost of our water and its conveyance.

Conveyance Costs

The largest cost variable in this plan is conveyance and its energy demands. As a rule, the cost and complexities of conveyance will always be the most limiting factor in selling and moving bulk water to markets. The highest cost variable within water conveyance is the uncontrollable cost of energy – even if purchasing 'water futures.' In our cost projections we always identify the cost of energy so that investors clearly understand this variable and its inherent instability over the life of 30-year minimum delivery contracts and its conveyance infrastructure. But we can now deliver our water to markets in China for about seven cents per gallon. Compared to the retail price of lesser quality local waters, selling at between $2 to $5 a gallon in these markets, this is an exceptionally good margin for our market partners.

Large Tanker Transport

The most efficient conveyance for bulk water from southeast Alaska to our targeted markets is large tankers that have been designed and built specifically for freshwater. We have, with global partners, explored the notion of tanker conversions (oil to water), but consistently found this option not economically viable due to the limited remaining life of these vessels after upgrade. Our preference is purpose-built freshwater V Maxx VLCC class tankers. Our new vessel designs allow greater access to sources, reduce turn area, increase vessel speed, decrease energy costs, and we maintain the freshwater load so that it is <u>immediately drinkable</u> upon arrival (no treatment needed). The result of this work is now paying off.

Middle Eastern Market

The Middle Eastern markets present a specific cost challenge to our Alaskan sources in shipping time/cost. Closer alternative sources in the Mediterranean we have worked on before may be far more cost/profit attractive, although not carry the "Alaskan" or "Glacial" marketing value but they are 'natural waters'. If a Panama Canal transit is preferred, it requires the use of inefficient small Panamax class tankers, causing significant challenges to market partners in cost per unit. We do not see the margins working within the price/cost in these markets now but are confident that will change. For example, in addition to the $250,000 to $500,000 fees round trip (RT) now charged by the Panama Canal Authority (not a controllable) is the added cost of tankers sitting idle for up to three days, even a week, before they can even enter the canal.

Security

Even if we go around Singapore with an Aframax, Suezmax, or VLCC class tanker the costs to conveying Alaskan *natural* water for Middle Eastern markets are significant per gallon compared to closer source alternatives and certainly far greater cost per unit then good RO desalination can produce. The most important concern with all of these 'other sources' is security – we are not allowed to own them – we have tried in many areas of the Atlantic. We have learned the hard way that source security is fundamental to a secure responsible water investment. The *natural* Alaskan glacial freshwater is uniquely secure and has a growing emotional consumer demand that, with our marketing strategies, will get more and more rabid.

The V Maxx VLCC tanker has the capacity to pump between 14/15,000 tons or almost 3M gallons an hour. With 90M gallons in the tanker this requires at least 30+ hours provided there are adequate (in market) pipes and pressure **at tidewater**. This is important as the longer a tanker sits in port the higher cost per round trip. Tankers are most efficient when fully loaded and traveling at max speed, not sitting in port. There are options such as using Water Transport Bags (WTB) that we developed for the Mediterranean markets for in sea bulk storage, which we discuss later.

Vetted Market Partners

With the assistance of Euromonitor, we are now looking for vetted market partners in very targeted markets in the ten largest coastal cities in China. Our market partners must 'take' our water at tidewater and then either store it or use it. If they want to start and stay with WTBs that can be an alternative for in sea storage. Although we will be highly active in our marketing work for these market partners, we will not be involved in the operation or financing of their side of this venture. Our experience in foreign markets has proven this to be a solid much safer strategy. In-market-partners are much more equipped to deal with political issues there, then we are. With proven bulk water brokers, now being vetted, we are fully confident that we can secure delivery contracts sufficient to meet our goals starting by year three. (Why three years? It takes 23 months to build our preferred VMaxx tankers with the modifications we know are essential, once we sign a delivery contract with a broker or buyer and initiate the order.)

Separate Conveyance Company

Although we have spent years studying a range of conveyance technologies for freshwater including transoceanic, to include our successful development of Water Transport Bags (WTB) we tested and proved in the Mediterranean, we are not wedded to the notion that AQUOEUS International, Inc. must own and operate these systems. In fact, given our years of knowledge and experience, we think having a separate company own and operate our needed conveyance is a more solid business idea and we are open to it. We would like to be a part of that company because we have significant knowledge of a host of applicable technologies that will enable vessel speed, clean fresh tasting water at market, and the very image of a tanker when arriving at market is critical. It cannot look like an oil tanker.

Needed Capital

Although the function and purpose of this plan is to secure needed capital to complete our effort to secure and own our first twenty Alaskan sources, we also outline the opportunities in our targeted markets, selling a unique *natural* Alaskan glacial water, we will also open the door to other markets with other solutions so that potential investors in the initial phase of this venture understand other – even generational opportunities. The old question of, "Where do we start?" is the challenge in developing any global business with such wide variables in markets, products, conveyance, and technologies. But our approach here will be to get started with twenty **secure** (owned) high-quality *natural* Alaskan water sources establishing enormous "brand" credibility in our closest markets, and then open doors to a wider array of global opportunities. It also puts us in front of market demand with the time to respond with total source security.

High Quality Clean Water

The retail product for this unique enormously high-quality *natural* Alaskan glacial freshwater under this Business Plan, is not just bottled water/beverages at the high end of each regions bottled water/beverage market. Additionally, the development of "Health Waters" targeted to specific consumers is a niche market in these regions that should be exploited in their specific markets as it can very successfully take advantage of the marketing platform developed for our water in these regional markets. Further we now have direct intelligence on several manufacturers who are currently paying exceedingly high rates for local filtered water and its essential filtration because it is contaminated and very costly to clean. Our water will not only be cleaner it will provide a base brand for these manufactured goods that significantly raises their value in their long-established global markets. AQUEOUS International, Inc. with its market research and sales development partners and team who support this venture bringing unique insights, technologies, relationships, and product line concepts that can <u>immediately</u> take advantage of these and other new quickly developing product opportunities.

Update NOTE:

Male sperm count continues to decline globally. This growing international concern is due to the increasing amount of estrogen (discharged mainly by women using birth control pills in their urine) in surface and ground waters in almost every major city in the world sets up a new angle for marketing our pristine product. Updated reports on the dramatic drop in sperm potency is already causing concern. Estrogen is also causing radical biological changes in all fresh and saltwater contained wildlife including fish and mammals.

Out of 20 different brands of mineral water available in Germany. They found that 12 of the 20 brands (60 percent) had estrogen contamination. In many cultures there is no more compelling 'fear' then the loss of male potency. Even the most advanced filtration systems, such as has been developed in Germany, can only reduce these hormones to a point – and remove essentially all-natural minerals. Again, a solid new marketing opportunity.

Personal: I only drink water from a 32,000-year-old glacier in Anchorage, AK unless I'm traveling. No estrogen there.

It's time!

Drinking Water: A Global Context

- ✓ 663 million people in the world live without access to clean water
- ✓ 2,200 children under the age of 5 die <u>every day</u> due to unclean water
- ✓ 1 child dies every 39 seconds due to unclean water
- ✓ 40 billion hours are spent yearly by women to collect water

Figure 2   Drinking Water: A Global Context

2. BACKGROUND

2.1 History of the bulk water business

Early Possibilities

Although there have been bulk water sales, including cross-ocean contracts for generations, the scope and context of this emerging global commodities industry really did not get underway until the early 1990's with the release of a paper, *Alaska's Bulk Water Exports*, on the internet by the Alaska Director of Water and Chief of the Alaska Hydrologic Survey. Unknowingly to the author, this paper for the first time attempted to examine the economic, political, and ecologic possibilities and then outline an approach. This paper was initiated because of several pending applications by California companies to begin exporting large quantities of freshwater from Alaska, initially to Southern California, as a backstop to pressing drought stress, incompetent state/local resource management and water subsidies, in those high demand markets.

With the release of this paper, several global and regional ventures in many areas of the world were initiated using a variety of approaches and technologies. Most notably was the advance of Water Transport Bags (WTB) in large scales by World Water, SA (a global consortium of three large international companies including AQUEOUS International) and Chaired by our founder.

Flaws in the Business Model

It was during this surge in the water business that it became clear that there were a few fatal flaws in the general business model. The failure to clearly understand the inherent insecurity of foreign sources and even government sales contracts, especially when purchasing/selling a commodity like water, was the leading and key issue. In nations, other than the United States of America, governments believe they have the absolute right to change any government contract or permission/permit with any change in government especially if such contracts deal with public necessities. Essentially, they believe that one government cannot hold a future government hostage to such 'contractual' agreements, especially in the sale or import of resources as critical as water. In some cases, national or local governments who sign such agreements change regularly, sometimes more than once a year. In another corner, although retained by a national government to help them look at the application of WTBs for some of their communities, we were not informed that all bulk water conveyance and sales were totally controlled by organized crime. At our next contract for another national government, we asked that question up front, and again found bulk water controlled by organized crime. So, it became clear that there was a need to create more certainty or source/cost security in any such agreements, especially with foreign governments who wish to "sell" their water and may be unwilling to 'deal' with the character of current owners/operators.

Understand Water

In several countries bulk water deliveries continue but at a vastly smaller scale and much higher cost/price then what we proposed. This is the case in Italy, Turkey, Cyprus, many of the Greek Islands, and parts of northern Africa and the Middle East. Even Japan, Norway, and Chile have explored the development and deployment of bag technologies. In addition, some of the largest shipping companies in the world seriously looked at the conveyance potential of this commodity starting with converted single hull tankers and looking at WTBs. For the most part they have not moved forward because of some of the limitations listed above, but also because **they do not understand water** and its wide range of uses and markets and range of value in them. Nor did they clearly understand or appreciate the critical timing of such a venture. Further, there are important barriers to entry into this global market that require knowledgeable, financially capable investors and market partners, willing to take the first essential steps and then allow the time for the venture to mature so that economies and persons have a high level of confidence in these notions. It is important to understand that water is often like a drug. Once economies are dependent on a water source, they are affectively addicted to it and its cost and delivery stability.

Security

The security of any quality water source is the paramount concern in this business and is often the second question asked by a potential buyer, second only to "How much?" If a delivery contract is signed that spans a minimum of 30 years, some as many as 99 years with negotiable 5 year adjustments. The buyer/market and its economic life is thus directly and enormously dependent on that water and its security and dependable delivery. Interruptions in conveyance, be they the result of man or nature, are not easily tolerated as the economic implications in markets are serious and almost immediate when water runs out. This is also why we strongly advise clients and buyers, especially governments, to diversify and expand their perspective and consider a variety of water sources and technologies as they plan their economic future and its security. Water supply diversity is becoming a mandate now in almost all national, city, or basin water planning – we believe in some part due to work we have done with many of these countries. It is what we have been preaching for over two decades.

Secure US Sources

This venture starts with a very secure (we own them as a property right uniquely protected by Alaska law and by the U.S. Constitution) group of sources in southeast Alaska, thus these concerns are minimized to the greatest extent practicable. A contract in the United States is still a contract enforceable in court and under the rule of law. In many countries contracts or agreements are not as secure, especially if it is with a government whose leadership changes regularly. Even within the European Union (EU) this is the case, and we anticipate this instability to continue without any modifications based on our discussions with the US State Dept and the US Dept of Commerce and our experience in the Mediterranean.

Secure Market Partnerships

There are ways to deal with this problem, but it requires <u>very secure market partnerships</u> – yes, within each market and their significant commitments to political relationships that can change based on factors that have nothing to do with water or the performance of this company. It has been our experience that the "under table payments" so often demanded by government officials after the contract negotiations are complete, have killed many a potential delivery contract. **We will not function outside of the law.** For any American company to pay such demands is a felony in America – for each payment, no matter in what country. The development of **extraordinarily strong local market partners** within each major market <u>**is essential**</u> to overcoming this uncontrollable cost and legal variable. Again, we believe it essential that we

19

have a **market partner/investor for every major market** regardless of type, purpose, function we enter.

Minimizing Conveyance Costs
This is the reason we have developed, and will present in this plan, business models in the purchase, conveyance, and sale of bulk water from Alaska to global markets. One model requires the seller to provide the conveyance and another model requires the buyer to provide conveyance be it owned or leased. The reason for these models is fundamental to who will take the conveyance cost risk in a minimum 30-year Delivery Contract with a foreign buyer. Again, we support a partnership with a third party in the bulk conveyance industry, rather than absorbing this single function within AQUEOUS. We are water and marketing experts, not conveyance experts – although we have learned a great deal.

It is important to understand that water is not just another commodity. Water is often a **religious** or deeply cultural substance in many markets. Water is **always** an emotional subject especially if you are "taking" it from a country or a people and "selling" it to another country or people. Even when those who think they own sources of water clearly understand the water being purchased is in significant excess to their current or future needs (which we always protect), there are deep emotional and thus political concerns. Working through these challenges in a wide variety of cultures and political systems requires extraordinary understanding, skill, sophistication, knowledge, persistence, and lots of patience. Relationships must be carefully constructed with each source and market that works in the best interest of this company over the next 50 plus years.

Market Volatility
Finally, to show just how rapidly markets and water values can change, in the late 1990s the State of California authorized its new Water Bank to begin acquiring water from sellers (mainly agriculture with federally subsidized water) within the state at $250 per acre foot per year which is equal to 1,234 cubic meters or 325,851 gallons. At that time this was a low but acceptable price for farmers to annually sell water they did not plan to use that year to the State for ready cash. By 2009 farmers were selling this same water at $5,500 per acre foot per year. And do not let me forget to again mention that in many cases this water, provided to farmers, is federally subsidized. This

escalation in price in just ten years is even beyond what we estimated, and the inflation continues. Scarcity increases demand and with the general political incompetence we have witnessed in California, it most likely will not change. But there is more: California farmers pay an average of only $70 per acre-foot for (again, most often federally subsidized) water to irrigate their crops. But they have learned that if they buy a $700 reverse osmosis water purification system, set it up in their barn, run 326,000 gallons of water (one-acre foot) through it; bottle it, the 'shelf' value of that acre-foot is suddenly $2.4 million. And that's water that is not unique in any way.

False Economy
During deep draughts in the last decade, Los Angeles announced a mandate to reduce its total consumption of water by 17% immediately. That is an enormous reduction in water consumption, and this came after most commercial users had already reduced their consumption by 20% to adjust price, in one of the largest economies on this planet. Again, from our perspective their problem is not charging 'actual cost' per gallon but continuing enormous subsidies that undergird and maintain a very vulnerable false economy at every level.

2.2   History of this endeavor in South East Alaska

Sitka, Alaska Water Sources
One of the first sources we explored was the one in Sitka, Alaska. With the old contract between the City/Borough of Sitka and True Alaska Bottling (TAB), terminated on December 8, 2009 and the cure date of January 21, 2010 then passed, Sitka extended a revised contract with TAB for one year, with new conditions. The amount of vessel lift required to continue the new contract was then 50 million gallons, the exclusivity had been removed, etc. This still left about 6.5 billion gallons a year available for contract to another buyer and whichever buyer shows up first gets the water they are willing to pay for.

We need to disclose that Sitka received the first bulk export permit from the state at the suggestion of our Founding Chairman, Ric Davidge during a regional economic development conference at which he spoke soon after he left state employment as Alaska's Director of Water.

Option for Sitka

AQUEOUS was then in position to offer the City and Borough of Sitka an alternative to TAB or any other potential competitor. That alternative was going to be AQUEOUS Global Holdings, SA as proposed at that time with a Middle East investor, and we needed sufficient credibility to warrant and leverage a decision in support of our Letter of Intent and Proposal before both the Sitka Economic Development Association Board (SEDA), who would make the recommendation to the local government, and the City/Borough of Sitka Assembly. We formally submitted a Letter of Intent and then entered quite discussions with the agent for both public entities in the further development of our proposal.

Although True Alaska Bottling (TAB) had (2006) attempted to move water under their contract with the City and Borough of Sitka they had not shipped even one gallon even today, although they often claimed they had moved water or soon would in their media marketing or internet postings. They instead made payments to Sitka in lieu of shipping, to retain their first rights under their original contract, but that contract ended December 8th of 2009. The latest extension required an additional payment of $100,000 plus all bills being up to date – about $120,000. This was paid, so their contract extension was then in place for 12 more months. They never moved water to any market.

TAB Extension

Although we were advised that both the SEDA and the City and Borough of Sitka Assembly were not likely to renew the TAB contract, due to TAB's failures to perform after many promises, consistent with the terms of their contract – TAB was the only formal proposal before the City/Borough of Sitka and an additional $120,000 was just too attractive to this small town government in there down economy.

Because of the TAB extension, the Sitka Economic Development Association and the local city and borough government remained open to the export of bulk water from Sitka to anyone who made a formal proposal and the required down payment ($250,000) for their first shipment.

It was our strategy to get our toe under the tent before any other new proposals were submitted for the remaining 6.5 billion gallons a year. Our ongoing contacts with the agent for Sitka responsible for the export of bulk water, suggested that we would be successful in this strategy.

Access Problems
This was an extraordinary opportunity for a company to step in and capture this unique high quality natural glacially influenced Alaskan water for bulk export with potential of 9.5 billion gallons a year. But it was our field work and onsite consulting with our tanker captain when we discovered a problem. Access with a vessel of sufficient lift capacity to make it financially viable – **did not exist** and would most likely never exist due to cost and the geographical layout of the area going in and out from the Pacific access. Again, we always found it good to do your own fieldwork.

Yes again, we learned the hard way that we must get our head out of the water long enough to see the shoreline. We wrongly assumed that Sika was accessible by a tanker of scale adequate to make the venture work. It was not and likely never will be. We later learned that they knew this and did not disclose it to us or any other potential buyers.

We are very often asked by Wall Street firms, the Security Exchange Commission, and many others, even internationally, about the Sitka source and their offer to sell bulk water. We have received calls from investment groups that included serious folks, only to have to inform them that the conveyance scale and physical access was the problem. Unfortunately, Sitka continues to try to sell, not in any way misrepresenting anything, other than to inform a prospective buyer that they cannot bring even a midsized tanker in to load. Various options have been explored, including the use of WTBs out to tankers, but again cost is the concern. It just does not pencil. Too bad, but again lesson learned. Maturity takes time and lots of learning from mistakes.

2.3 History of AQUEOUS International, Inc.

1991
The ideas that remain the basis for creating AQUEOUS International, Inc. were born while Ric Davidge, MPA/PM was serving as Alaska's first State Director of Water and Chief of the Alaska Hydrologic Survey. He accepted

this position, after saying no to the Governor twice, but left the private sector at the insistence of Governor Hickel (a friend) who had long been interested in the development of what was then an idea for a water pipeline to California. Obviously concerned with potential political opposition, the Governor specifically charged Mr. Davidge with the assessment of understanding the ecological, economic, and political viability of such a pipeline. The Governor told the press that the position was essential to Alaska's future and that Davidge was qualified for this task because of his former positions in the U.S. Department of Interior by appointment of President Reagan; his work with the Grace Commission on behalf of the Department of Interior; as the Federal Commissioner to New Jersey (while a member of the Reagan Team); his diverse work for the U.S. Senate resulting in a number of federal investigations; his history in Alaska most recently leading all environmental compliance efforts in response to the Exxon Valdez tanker still, as well as his proven general public and private management and project management skills.

After 30 days Davidge reported to the Governor that the pipeline idea was ecologically possible but not economically or politically probable given loud opposition by British Columbia, Washington state, and Oregon (offshore submerged pipeline opposition), most from the national Green organizations. The governor was not pleased, but Davidge has a reputation of being incredibly open and honest in his public and private work and with the Governor in a previous function. The pipeline idea was set aside.

However, during these first 30 days as Alaska's Director of Water, Davidge discovered that there were several applications from a California company to take and ship bulk water from Southeast Alaska to California. The California company had been rejected by the new Green Party in British Columbia, Canada even though they were under contract for bulk exports prior to the election that changed that Province's government. He also discovered that there was no legal framework for Alaska to review and approve such requests let alone require any payment to the State for the water it would permit for bulk export. Almost all water, including ground water in Alaska, and including ice, snow, and glaciers are owned by the state. Davidge quickly prepared emergency legislation that was endorsed by the Governor and an emergency joint House/Senate Resources Committee meeting was called. This resulted in the first substantive bulk water export law in the world, that considered both the ecological and economic issues and resulted in setting a global standard for the 'use it or lose it' (w/in 5 years) limitation to help keep 'speculators' from tying up Alaska's potential water export market.

1993 Paper

After the passage of Alaska's bulk water export law, Davidge announced that he would prepare a paper on these two, often opposing issues - the economic and the ecologic viability of a new generational industry for Southeast Alaska. In 1993 this paper was sent to friends in Congress and a technical review was requested allowing for an independent assessment of the various costs and impacts. The response was positive, and Davidge released the report on the internet that year. Only to soon realize, that some of the base conveyance cost numbers he had been given - were wrong. He made that announcement at an international meeting in Japan later.

"Water CZAR" Davidge

As the Alaska representative to the President's Commission on Western Water Policy, and as a member of the Western State's Water Policy Council, Davidge was quickly thrust into not only national but international attention on these and other water related issues. He was soon dubbed 'The Water Czar' by several state, national, and international media. He was also a featured presenter in the documentary film, 'Blue Gold' and an early presenter at the new World Water Congress and other international groups on his ideas of a bulk water industry and the eventual global commodification of '*natural*' freshwater. He also became an outspoken critic of desalination, both for its political behavior especially with small local governments needing water, their marketing misrepresentations, and the short (pregnancy/early childhood) and long-term serious health impacts of demineralized water. He continues this effort.

Davidge then began pointing out the growth of hormonal and prescription medications, generally now referred to as 'pharma' in the sea and in most large cities' ground and surface waters, and in water recycling systems. The recent discovery of several pharmaceuticals and their yet understood interactions in vegetables watered by toilet-to-tap water recycling systems has confirmed

Davidge's concerns. His papers on the growth of these contaminates in freshwater, both surface and groundwater, are often requested globally. He continues to monitor the development of new technologies via patent reviews that might someday provide safe and healthy recycled water that is economically viable for direct human consumption – but to date there are <u>none</u> regardless of what they may claim.

Upon his departure from government as Alaska's Director of Water and Chief of the Hydrologic Survey (he had only committed to serve in this position for 2 years) he accepted a temporary role to lead several challenging initiatives for the Governor, leaving state government in 1994 to return to the private sector.

This background is offered to help you understand how far these ideas have come. Bulk water exports of some scale have been underway for generations, but no one conceived 'natural' freshwater as a globally traded commodity nor an industry at the scale suggested by Davidge as early as 1993.

Although Davidge returned to the private sector, he was constantly called on to participate in a wide variety of freshwater meetings and gatherings around the world. Thus, he formed Alaska Water Exports, Inc. to enable travel and consulting opportunities. He continued his work on the two questions of ecological and economic viability with very specific interest in SE Alaska but also in other areas of the world and his constant assessment of the growing pharma contamination and health limitations of demineralized water.

1994
In response to the growing international demands on the company, Alaska Water Exports, Inc. changed its name to AQUEOUS International, Inc. and continued international consulting work for a wide range of companies, nations, cities, and international security groups most of which required strong confidentiality agreements. He was also contracted by the Joint Canada/USA Boarder Commission to participate in a summit to review proposals for bulk water exports from the Great Lakes. At the end of these meetings, Mr. Davidge provided comments on the proposed project and its impacts on both Canadian and American cities that use water from the Great Lakes. The mayor of Chicago (Richard Daley, five times reelected) was not pleased that

Davidge called for American cities to meter this freshwater uses at least as well as Canadian cities. Years later, while doing work for the nation of Albania, in a visit with the American ambassador there, the lawyer for Mayor Daley at the time of the meeting, joined the discussions on Albania's water future. Small world.

Davidge then began pointing out the vulnerability of static physical plants to both natural and manmade threats in testimony and speeches globally. This immediately brought in consultation demands that had long term impacts on the company that further pushed it into many international (secret) contracts, especially where source, conveyance, storage, or processing water security was a serious issue. Consistent with many of his contracts, he was and remains severely constrained in what he can say or reveal including even the name of his clients. This demand dramatically expanded following the terrorist attack of September 11, 2001. Davidge can however discuss the border questions of water security and resupply but not specifics to water security issues to nations, companies, or 'interests' he has advised.

2000
While assisting Mayor Sarah Palin (later nominated for Vice President of the United States) to outline and implement an aggressive economic development program for Wasilla, Alaska, Davidge received a call from a representative of NYK Japan on behalf of several global investors interested in 'considering' entry into a global freshwater industry/market. Taking a leave of absence, Davidge participated in a large private (invitation only) investors conference in Tokyo that reviewed all aspects of the global freshwater crisis, all the old ideas, and any potential new ideas or technologies that might help reduce cost and open new sources and markets. Following a series of national/international experts on these issues, Davidge was the last speaker and was introduced with context to his 1993 paper. He abruptly announced in his opening comments that the cost/income numbers in that paper were wrong. This caused an immediate stir but reinforced his background of candid honesty in such matters.

The information that had been provided by the tanker industry was not correct and new tanker designs were underway that could likely change everything. He also introduced a new conveyance technology, Water Transport Bags (WTB) that were finally being tested for open sea conveyance in the Mediterranean.

During Q&A those invited challenged Davidge on several points such as the ecological, political, and economic viability of this possible new global industry and his opposition to the recently adopted "water is a human right" notion of the United Nations. He backed this up with a review of nations fully implementing market-based distribution systems, such as Chile, and how successful they have been compared to the failed government regulatory approaches applied mainly in the United States. After the conference, Davidge was asked to join a small group at the NYK board conference room to include a Saudi group and others. This meeting went on for hours ending with many open questions.

Just days after returning to his work in Alaska, Davidge received a call from Saudi Arabia and was asked to meet a group in Paris for an organizational discussion on the formation of a new global water company. His ticket was already at the airport, and he had to leave the next morning, incredibly early. In Paris, after a day's discussion, World Water, SA was formed to include NYK Japan, ALJ (a large Saudi group), NWS a Norwegian company, and Alaska Water Exports, Inc. Davidge agreed to be Chairman of the venture. The purpose of this venture was to aggressively explore the development and use of WTBs starting in the Mediterranean, and to keep aggressively moving forward with new conveyance concepts, identifying sources, and markets. Davidge returned to Alaska and advised Mayor Palin that he would have to resign her team to immediately work with this new venture. This was a very timely development.

WTBs
World Water, SA needed an extraordinary partner in the further development of WTB's with concern to its operative life and functionality. Albany International Research Co was a wholly owned and independent contract research house specializing in industrial fabrics, fibers, and coatings. Areas of activity and expertise were broad and deep, and included exclusive NASA space shuttle exteriors, hot process industry specialty insulation, fiber security features for currencies, fuel cell bodies, biomedical human parts, synthetic vascularized skin, and synthetic down, among others.

The general mission of Albany was to develop new technologies compatible with the corporate core strengths and translate these into new products and business lines. The research strategy was a self-funding endeavor. Today, Albany's largest noncore business line involves composites and other unique offerings for aerospace as a spin-off from these efforts. Albany was the perfect fit on the WTB initiative.

Ed Hahn

Ed Hahn led Albany International as its President, as well as functioning as the corporate Sr VP and CTO for all lines. During Ed's tenure, he and his teams were the Water Transport Bag development partner with and for Ric Davidge and World Water SA. Large scale models were designed, constructed, and tested, and many significant ideas were granted legal patent status. This major project led to a close working relationship and professional friendship between Ric and Ed. They have since collaborated for more than a decade on water related projects.

The development of WTBs was then, and continues to be, an interest of AQUEOUS International, Inc. even with its limitations because it offers unique cost savings and critical versatility that tankers and pipelines do not. Immediate offshore freshwater storage is just one example of how these bags can help, especially water startups, deal with the cost and security risks of freshwater storage on shore. Working with their partner, Albany International, a new patented fabric was developed for WTBs that greatly extended the life of the bag and its flexible strength during tow. AQUEOUS International, Inc. is still working with sources and markets in many parts of the world on the use of WTBs for various applications. The project between Turkey and Cyprus (website video) clearly shows the functional and economic viability of WTBs. But there are limitations such as speed/energy cost, total scale (up to 50,000 cubic meters) and aft fishtailing at increased speeds. All these limitations have now been resolved by Ric Davidge and Albany; however, Albany has set the WTB program aside due to other pressing global priorities and what were perceived to be low demand at the time.

New Technologies

AQUEOUS has and continues to keep close eye on any new water-based technology with search engines that monitor any new patent applications that may be of interest to the company. In 2011 AQUEOUS was alerted to the testing of a new desalination system that looked very promising. After some investigation of a limited pilot, AQUEOUS and two investors became advocates of this new system and initiated several tests to prove its ecological and economic value. This was the company's primary focus for about five years given the global potential for this system in a wide range of water cleaning applications. But there were issues that continued to trouble AQUEOUS and finally an agreement was structured with a respected 'virtual pilot' engineering company to 'prove' the inventors' claims. The results showed that in fact the system did what it was represented to do, remove all salt and most minerals from the sea water and convert that 'waste' to a powder of significant interest to various industries. But the energy demand originally represented by the inventor was found to be very wrong. In fact, the energy demand/cost was so exorbitant that it removed it from any serious market consideration for almost any application. Although this was a big disappointment to AQUEOUS, an enormous amount of new intelligence on desalination and other types of demineralization systems was gathered by the company pointing it back to 'natural' freshwater as the commodity of interest.

The interest by AQUEOUS in demineralized water changed again with the public release of the 1960's top secret Russian report on prolonged consumption of this type of water by humans and mammals. Davidge went back to be a strong critic of desalinization due to its impact on humans especially pregnant women and young children, livestock, and in some cases even agricultural irrigation.

2008 Mr. Davidge continued to monitor new water technologies and began to give greater attention to new tanker designs that might make a significant difference in their 'day cost'. He continues to speak at national and international venues on the probability of an international bulk water market. Water continues to be his first and main interest, along with veterans and national/state political issues. He has also written successful political nonfiction books rated 5 stars on Amazon.

2019- Today AQUEOUS International, Inc. has several projects underway including the acquisition of almost 100 Billion gallons a year of exceptionally high quality '*natural*' glacial freshwater from identified, assessed, accessible sources in SE Alaska for bulk export: and the development of several new very high-quality freshwater sources in the South Pacific, Eastern Russia, Albania, the Adriatic Sea, etc. The development of vetted 'market partners' in several Pacific and Atlantic markets continues with its international market research partner. The current plan is for these exports to begin in 2023 to include the use of a new class of 'freshwater' tanker at 90 million gallons a week per source with a delivery cost of under $.07 per gallon to China as example.

Foreign Investment Interest Declined

Although there have been several expressed interests by foreign people and companies to invest in AQUEOUS, the team has decided, and annually confirms, that AQUEOUS International, Inc. will remain a majority owned Alaskan based company. In 2019 AQUEOUS found a strong Alaskan investor and is in the development of partnerships with high quality freshwater sources and market partners around the world. Then we got COVID and things stopped, especially our planned trips to China.

2020 Got a call from a group in Canada who had been looking at the bulk water business for some time. They wanted to play but had a long list of questions that were answered. It became clear that what they really wanted was our work, without payment.

AQUEOUS will continue to provide **highly confidential** consulting to companies and nations concerned with their macro freshwater strategies and security in the face of natural and manmade threats. The loss of a major desalination plant, for example, or even a water treatment plant or storage facility can shut down an economy for many months causing dramatic economic and political convulsions. In some circumstances, such interruptions cannot be recovered in a year.

Key Elements of the AQUEOUS Team

As the AQUEOUS business concept has matured, a significant world class team of likeminded competent global professionals have become closely allied with our vision of a high valued '*natural*' freshwater supply in bulk. Water experts with broad contacts and decades long relations, financial and operational modelers, major operational leaders, a world class mega class ship captain/pilot with Arctic and Antarctic experience, Asian market financiers with global generational active connections with sovereigns and large-scale industrials, and official nation representatives with both sources and markets are all very connected with the efforts of AQUEOUS - all are key elements of the team.

AQUEOUS International, Inc. can be found at www.aqueoususa.com (907) 229-5328 or email Ric Davidge directly at rdavidge@aqueoususa.com or ricdavidge1@gmail.com .

3. Rationale for this venture

Why is it justified, and we believe necessary, to ship bulk *natural* freshwater, on a global basis, for drinking or manufacturing or possibly municipal 'mix' water?

Bottled Water - best indicator of a water market's quality and quantity demand

Bottled water is, in most of the markets we are targeting, the **only** means for consumers to acquire a water source for personal consumption that has any quality integrity. Unlike most (not all) municipal waters in the United States, the public water sources in China and India are not fit for human consumption. You will also read in this plan about recent studies evaluating the quality of the bottled waters sold in Asian markets and how contaminated they are. Bottled water is THE source of water, of various qualities – few are natural- for people to drink in China, India, and most countries across the planet who are growing more and more aware of the extraordinary contamination of their local (surface and ground) waters.

Bottled water sales in the Asia/Middle East region more than doubled and then doubled again in the past decade. Growth has been spectacular in several countries, with bottled water fast becoming the norm for in-home, in-office, and on the move hydration. But none of them have our brand strength.

Opportunities and Challenges
Both regions offer differing opportunities and challenges. Asian markets are often young and emerging, with evolving distribution structures and a rapid influx of new players. There is also great potential for sustained growth in countries such as India and Pakistan, which have large populations but low bottled water penetration. The Middle East, though with some new emerging markets, is generally characterized by a mature and well-established bottled water market. Moving from ingestion of desalinated/demineralized water to a packaged product is key in the future of our *natural* water. But again, and importantly they do not have our brand – pristine sustainable secure Alaskan

natural glacial water from sources as old as 32,000 years.

Although there are a wide variety of ways to "solve" a freshwater shortage in almost any market, the questions are always: water quality, cost, security/reliability of source, distribution capability/cost, *natural* vs manufactured water, and for how long? These questions also apply to the concept of bulk water conveyance for industry and municipal mix*.

Municipal mix means Alaska water mixed with local municipal water for distribution.

Bulk freshwater conveyance, as a concept, is well established in many regions of the world. Large shipments of freshwater daily transit the seas between Turkey and Cyprus. Similarly, a variety of weekly shipments reach a growing number of Greek and Italian islands. Remember that many of these are controlled by organized crime. Many NATO facilities in the Mediterranean are served regularly with freshwater by ship, bag, or barge. These are just a few examples: but the scale of these freshwater conveyances is small compared to the need of most major markets, the capability and security of unique high value *natural* water sources, and the diversity of conveyance technologies including our WTBs are now available, where those numbers work.

To our knowledge, no one is delivering water specifically for bottled water, beverage markets, or industry with the unique qualities of the Alaskan *natural* glacial water or at the volumes anticipated by this venture to **any** market.

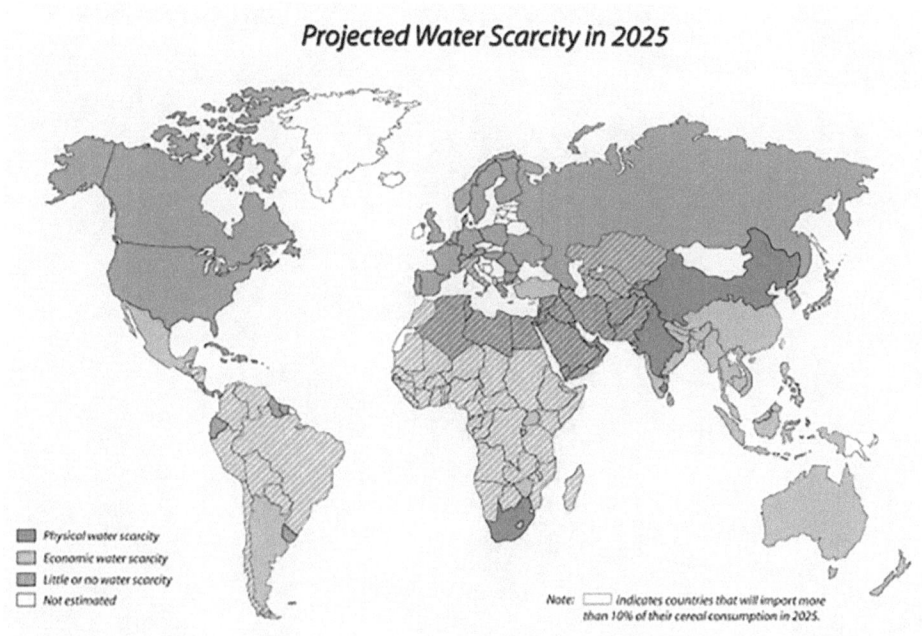

Figure 3: Projected Water Scarcity

Survey Error

Again, we point out the error in this survey. They **did not consider** the extent nor the malignant advance of saltwater intrusion/contamination in the coastal aquifers of dozens of countries shown here as not concerned with serious freshwater scarcity. But this image does help most better understand where the pressure is for more freshwater. Desalination will not solve this need, as more people become aware of its harm (few minerals) to infants and others and its huge discharge of brine now collecting and raising RO costs by as much as 50%. Further the brine continues to harm almost any ocean life.

Ric Davidge, MPA

4. TARGETED MARKETS

'First, know your markets.'
We do not look only at bottled water markets but use them as an exceptionally good indicator of consumer demand for premium valued water, as we provide. We also look at industries who have shown an interest in bulk imports because the cost of their filtration of provided local water is at the point that importing Alaskan natural glacial water is within their margins. We have at least two manufacturing groups in China extremely interested in bulk imports, with various specifics. Again, working through South Korea offers real opportunities to meet these Chinese market demands without the corruption or political intrusion we know of in China.

Why Mexico? The fun reason is we get to wave at California as we ship down the Pacific shoreline. We accomplish this without conflict with the Jones Act and the hope that a change in CA governance might enable a joint effort to get a congressional exemption for bulk water into southern California – a huge market.

Why China, then India, and then the Atlantic and the Middle East?
China is now the world leader in the consumption of bottled water. The top ten cities in China with tens of millions of residents, all with large port access, and represent only part of this markets fast growing demand for clean natural freshwater.

In addition, huge markets in India, the Middle East, Northern and Southern Africa, and even portions of Europe, Southern California, regions of Mexico, etc. await the import or creation of "new water" so that they <u>and their economies</u> can survive and grow. The more water consumed via bottles reduces the demand for municipal water for human consumption and opens the possibility of redesigning municipal water delivery systems. It is important to keep in mind that bottled water includes all sizes, from personal "pet" bottles at half a liter to 1.5 liter, a gallon, 2.5 gallon, and even the five-gallon water cooler standard often found in offices but now more and more found in homes.

4.1 China

A National Market Perspective: Even despite the global economic gloom earlier this decade, the Chinese bottled water market is still growing and maturing quickly with an increasing consumer awareness of, and demand for "quality" and a deep uneasiness about locally sourced beverages due to formerly unknown access to current national market reports. As a part of this growth and market maturation, consumers are coming to understand and seriously appreciate the real scarcity of particularly good reliable *natural* water from local surface and groundwater sources, **even when filtered and bottled**. We saw this trend as predictable well over a decade ago based on our experience in China and in other markets as all trends were moving in our favor. But the rate of growth was greater than we anticipated. Why?

Some context: 2013 early am – local media report:
- "Thousands of dead pigs in a Shanghai river have cast a spotlight on China's poorly regulated farm production, with the country's favorite meat joining a long list of food scares.
- As of Friday, the number of carcasses recovered in recent days from the Huangpu river -- which cuts through the commercial hub of Shanghai and supplies over 20 percent of its drinking water -- had reached more than 7,500.
- Shanghai has blamed the farmers of Jiaxing in neighboring Zhejiang province for casting pigs thought to have died of disease into the river upstream, but officials from the area have admitted to only a single producer doing so.
- The city has stepped up inspections of markets to stop meat from the dead animals from reaching dining tables of its 23 million people."

Again, this report is from a local paper in China as they watched the pigs bob along past Shanghai homes.

Figure 4: Dead Infected Pigs litter Shanghai Beach.

This was only the beginning as the total number of dead **infected pigs** in the principal water source for Shanghai, eventually each a total of over 16,000 dead pigs in their water. Hard to believe, but if you travel in most undeveloped countries, and China is no longer 'undeveloped' you see things like this often. Agriculture has dumped everything into the rivers and streams and the ocean for thousands of years. It is a part of the culture and an accepted behavior. In many counties, including Europe, the only solid waste disposal concept is dumping it into even dry streams in expectation that the spring thaw will flush it out to the sea.

Although this is shocking, it is not rare. Certainly, the numbers are, but the problems of surface and groundwater contamination have long been a red face to China. They have responsible laws and regulations, but the corruption at the service level is a serious concern - when there is any local enforcement.

2018

Almost 90% of all groundwater in Chinese cities were affected by fecal coli (human and livestock) and industrial and agricultural pollution as well as 70% of China's rivers and lakes. There have been a high number of river pollution incidents in recent years in China, such as the drinking water source pollution by algae in Lake Tai, Wuxi.

Because of these shocking nauseating incidents along with economic growth and wealth accumulation and thus personal affluence in the major cities from north to south all withstanding massive immigration from the rural agricultural areas to growing city affluence, and don't forget television and social media – yes, it is controlled by government but much still gets through, to tens of millions of Chinese multiple times a day who reach for bottled water as the **only** source of what they believe to be real clean safe drinking water. And the growth of imported waters continues to grow with international distributors flooding their products on shelves. The shelf cost of importing packaged water is extremely high. **But no one has our brand**.

Poor Bottled Water Quality

According to the International Monetary Fund (IMF), China's output was forecasted to grow at 6.7% in '09 and 8% in 2010, it did. But in 2017 things began to change i.e. the election of President Trump and his international trade renegotiation agreements – have not harmed the steady growth of bottled water as local awareness of terrible water contamination increased – even accelerated. But despite this economic wonder, the **quality** of tap water and most bottled waters/beverages manufactured in China using local sources still falls far behind western water quality standards – and the people now know it.

China Bottled Water Growth

In our recent four-week visit across China, we found a wide array of bottled beverages for sale at every stop, an opportunity not dissimilar to any other market in the world. China's bottled water industry has advanced quickly and achieved high volumes, despite extremely low but growing market penetration, especially in the larger cities. Volumes have surged from under eight billion liters in 2000 to 21 billion liters in '09. In 2013, China surpassed America, becoming the world's biggest bottled water market by volume. According to

China Daily, the market has steadily grown in the years from 2010 to 2015 from 19 billion to 37 billion liters of bottled water. It is expected to reach 49 million tons of total annual consumption by 2020. Market growth is anticipated to remain in the double figures for decades forward.

In 2008, it was estimated that 78% of the value in the Chinese bottled water market was attributable to still water in retail packs – nearly 59% in sizes up to and including one liter. That has now more than doubled and may double again in the next year or two.

Contamination

Although the national government of China recently reviewed water standards and compliance including local sources and manufactured beverages, they have continued to fail to improve the actual quality of tap water in most regions of the nation. Most experts believe the quality of tap water will continue to decline due to increasing pollution resulting from industrialization, population growth, agriculture, and mass migration, and the lack of any capable public sewage systems across most of China. The most often found contaminate in water in China is **fecal coliform** (both human and animal). This is **not** discussed openly in China, but it is the reality and most Chinese know this now that they have smart phones and the Net. This perception of low quality, if not danger, is one of, if not the key factor that has helped drive the growth in Chinese bottled water markets and is directly consistent with our market targets.

Figure 5: Pollution

Our target of mid to high-end bottled water and beverage products is precisely on point as this multi-billion consumer retail market is poised to buy better "**quality**" even remarkably high end 'fine' waters. When you add *'natural Alaskan glacial'* we passionately believe, based on our discussions in this market, that we can set a new standard and increase immediate demand on the shelf.

Their confidence in local water used by bottlers is constantly lowered with almost monthly stories of pesticide, industrial waste, and pharmaceutical contamination. The recent international scandal of contaminated water manufactured by Master Kong (brand), produced by Tingyi Holdings has further pushed bottled water consumers to **seek a higher quality of water source**. More and more Chinese consumers are now reading the labels on bottled water to ensure that the **source is pure**, rather than rely on local filtration. But there is no reference to *natural let alone glacial*. Price is still the leading factor in choice, but as has happened in so many other bottled water markets around the world, this is changing. Price is also perceived as an indication of quality – which has been the Evian strategy since its iinception in the early 1800's So, our source enables much higher prices than even imported French water – that has been pissed in for well over 2,000 years.

Figure 6: Pollution Clogged River.

Polluted Rivers

The above photo is not a sewer, but a river and the primary water source for a growing city in China. It is the standard, but there are significant other contaminates you do not see such as heavy metals and other industrial waste that does not float. China hides the number of deaths and serious illness such as cholera - the world's largest killer due to contaminated water. We continue to try and secure this data as we have been told it is staggering.

The point is that our brand will overtake any product (locally sourced or imported) in the market, not only for bottled water but all beverages that use our imported *natural* water as their base '<u>made from pristine natural Alaskan glacial water</u>'. We surpass any current product in China including Icelandic who keeps asking us for meetings in L.A. that we have turned down.

So, our point is that China is the strongest market, with accelerating demand, we can find in the pacific. It has the need, the emotional concern with contamination that they have personally witnessed, and they have the income to support a new line of mid to high-end beverages.

And there is more, much more

We have been contacted by industries in China about buying our water. It started with the cosmetic manufacturers and then we got calls from computer component manufacturers who all wanted to know what our price was. These calls were triggered by the Sitka scam. We did not disclose them but offered to keep them in mind as we develop. Their problem is not new in freshwater demand. It is now costing them way too much to clean the local water they now buy so that it does not harm their products. And then they must dispose of their waste which also is increasing their cost.

With growing consumer affluence, a rapidly emerging middle class – in fact the largest in the world, and the increase in foreign tourism in part due to the 2008 Beijing Olympics, China continues to present positive indicators for bottled water and beverage industries.

Marketing Help

Keep in mind that we not only provide the water, but it is also unique in this and most markets, we help them with our marketing. We have worked with Euromonitor International, with local offices in any market we consider, and certainly China (main office in Shanghai) who have studied the bottled water and beverage markets there for decades. One of the first things we do once funded is trigger two reports from Euromonitor on bottled water, beverages, and industrial opportunities globally – and the vetting of serious capable market partners.

An example of Euromonitor's knowledge and experience is a May 2009 Bottled Water Market Study with an attached interactive statistical database on what appear to be a wide range of product lines and regions across China. This and newer studies are available for ($900) immediately. We recommend these most current studies be acquired as soon as possible to assist in our business development and marketing.

Sophisticated Targeted Market Research

We have also initiated discussions and have a ND/NCA in place with Grail Research an exceptional market analysis company with offices in China and India. We have an initial proposal from them allowing us to do very sophisticated targeted market analysis. Once we review the Euromonitor studies and databases, we can decide if the Grail effort is necessary and/or how we want to target their market analysis.

With the assistance of a friend, we secured earlier this decade some initial market information that is very promising. This information has not been independently confirmed but based on its source and their 15 years of experience in the imported beverage (wine) industry into China, we believe these numbers are worthy of discussion.

BRAND	Jenny Lou's Caters to Foreigners		Lotte Mart Department Store with / Grocery Store	
	Yuan	USD	Yuan	USD
1.5L Domestic				
Coca Cola 冰露			1.95	$ 0.29
Nongfu 农夫	2.8	$ 0.41	2.5	$ 0.37
Nestle	3.5	$ 0.51	3.15	$ 0.46
L'Origen 蓝润	3.5	$ 0.51		$ -
Watson's	5.9	$ 0.87	3.9	$ 0.57
Tibet Spring			19.9	$ 2.92
1.5L Import				
Canada Icefield	9.9	$ 1.45		
Volvic	21.9	$ 3.21		

Figure 7:   Initial Market Information.

Euromonitor does this ongoing level of detailed research on retail products in specific segments of markets and in even broader markets around the globe. They constantly inquire: What is your targeted market?

Remember, we believe, based on our preliminary analysis using our choice of the VMaxx VLCC class tanker, that we can deliver pristine *natural* Alaskan glacial water to China at under $.05 per gallon. We believe this to be a very conservative number based on all the information we have to date. If our numbers are even wrong by half, we are in the market with an extremely attractive price for extraordinary mid to high end water with absolute premium top-of-mind marketing potential.

Danone currently dominates the Chinese bottled water market through its joint venture relationship with Hangzhou Wahaha – a partnership that accounted for more than 30% of the market value sales in 2006 – **that has soured dramatically since April 2007**. The following three-year dispute resulted in a series of lawsuits worldwide, they all ended in Wahaha's favor. Besides Wahaha, Danone also receives sizeable contributions to sales from Shanghai-based bulk water supplier Aquarius and business units Robust and Health, all of which the company has a stake in. Key international brands such as Evian and Volvic also add sales in China, in much smaller volumes.

Figure 8. Chinese Polution.

You think these pictures are bad. You should see it up close and personal. The stench is overwhelming.

Master Agreement with Coca-Cola

The Danone agra-food group, which controls 55% of the Shanghai markets for water services to homes and offices, has acquired 50% of Aquarius Water, a household water distribution company based in Shanghai. Aquarius Water posted a turnover of 170 million Yuans in 1999 (Euro 23.3 million). Of further interest is the announcement that the Coca-Cola Company and Groupe

Danone, the owner of the Evian brand and the world leader in the bottled water category, announced a Master Agreement under which Coca-Cola will manage marketing execution, sales, and distribution for the Evian brand in North America (U.S. and Canada). Evian distribution will continue through the Coca-Cola System, including Coca-Cola Enterprises. This sounds like a good sales contact for our mid to high-end natural Alaskan glacial water product.

Other multinationals have an important presence in the Chinese bottled water market such as Nestle with its internationally recognized brands Vittel, Perrier, S Pellegrino, Nestle Pure Life, and Nestle Aquarel. Due to the higher price point of these brands, which would be compare with our imported Alaskan natural glacial water, most of the contribution made in the country comes from key domestic brands. This is a KEY point of information. Using unique pristine natural Alaskan glacial water as the feed stock for a bottler in China allows that local manufacturer to take advantage of their existing bottling, distribution, and retail systems with a far higher quality water product with enormous brand/marketing value.

Coca-Cola has a share of the Chinese bottled water market, predominantly through its Ice Dew and Sensation brands. Coca-Cola has recently expressed an intention to increase its presence in China. In September of '08 Coca-Cola made a $2.4B bid to acquire the leading Chinese fruit juice giant Huiyuan Juice Group, the largest intended acquisition of a Chinese company by a foreign company. Six months later the bid was rejected by the Chinese Ministry of Commerce because they believed the acquisition would give the new company unfair advantage from its expanded market share. This was the first such prohibition ever issued by the Ministry under their anti-monopoly law.

Coca-Cola has said it will continue with plans to invest $2B over the next three years to expand bottling plants, distribution infrastructure and marketing across China. Coca-Cola has also recently opened a $90M Global Technology and Innovation Center in Shanghai. Currently Coca-Cola controls 50% of the Chinese carbonated soft drink market, which is now receding, and is well placed to develop other beverage choices.

PepsiCo relies on volume sales of Atlantis in China. They have plans to expand production, research and development, marketing efforts and enhance their sales force to build brand awareness among Chinese consumers. Their four-year plan will cost PepsiCo $1B and should help build sales in markets away from its home market to offset weakening sales in the United States. Nothing like a little motivation for the importers of pristine natural Alaskan glacial water.

Two of the most notable Chinese bottlers are Nongfu San Quan and Hong Kong based AS Watson. There are an estimated 1,500 local brands in China and their proliferation puts further pressure on competition as well as preventing market consolidation. Two thirds of the Chinese markets are accounted for by other smaller brands. Their scale and low prices pose challenges, but we are not introducing a low-end product. It's a different market.

4.2 China's Market Outlook

March 5, 2019   Chinese Chamber of Commerce report
Why Bottled Water Became Best Beverage in China

"Health concern is the number one reason for bottled water becoming the popular drink among all drinks in China. If food is gold, water is silver, only <u>healthy food</u> is not enough, must with healthy drink. Chinese government promotes healthy lifestyle for some years now, it achieved a glories result. Today, Chinese consumers are more sensitive to what they put into their mouths, they do take extra care about their health when linked to pay hospital bills, and rather go for healthy food and drinks than going to hospitals. Consequently, bottled mineral water is the popular common daily drink for old and young in China, with growing numbers of Chinese consumers are embracing a more active and healthier lifestyle, their beverages consumption habits are moving beyond soft drink, energy drink, and are getting on with healthier bottled water consumer products. Nowadays, most Chinese consumers consider bottled water is the best healthy beverages, among other alternatives, especially on the road. The average per capita consumption stands at 84.2 L in 2019; in global comparison, most revenue in the bottled water segment is generated in China (amounts to US$48,656m in 2019), according to Statista. Imported bottled mineral water is one of favorite imported beverages in China market; new lifestyle trends boost China bottled water imports and investment." (no edits made)

(standing by for new update from Euromonitor)

5. Our Top Ten Market Targets in China

5.1 Shanghai — 23.4 million people (first market target)

Figure 9: Shanghai.

- **Area:** 3,900 km² (1,500 sq. mi)
- **Density:** 6,000/km² (15,000 per sq. mi)
- **2018 GDP:** 3.27 trillion yuan (464 billion USD)
- **GDP per capita:** 140k yuan (20,000 USD)

Shanghai is **the undisputed most populous and wealthiest city in China**. With a name synonymous with world trade, Shanghai has the largest and busiest port in terms of containers and cargo tonnage, a grand business district, two large airports (Pudong and Hongqiao), the world's fastest train (the Maglev), and a network of elevated highways.

5.2 Beijing — 18.8 million people

Figure 10:   Beijing

- **Area:** 4,100 km² (1,600 sq. mi)
- **Density:** 4,500/km² (12,000 per sq. mi)
- **2018 GDP:** 3.03 trillion yuan (430 billion USD)
- **GDP per capita:** 160k yuan (23,000 USD)

Beijing is large primarily because it is China's capital. It is China's **largest city by area**, as Shanghai's urban population is more concentrated.

It is a political, educational, and cultural center, with light industries (science, technology, and research) dominating over mass manufacturing.

Beijing has the world's largest airport, and an extensive, mostly new subway system, but ongoing traffic congestion issues. The Forbidden City still features strongly in the core of its 6-ringroad concentric layout.

5.3 Tianjin — 12.8 million people

Figure 11: Tianjin.

- **Area:** 2,800 km² (1,100 sq. mi)
- **Density:** 4,600/km² (12,000 per sq. mi)
- **2018 GDP:** 1.88 trillion yuan (266 billion USD)
- **GDP per capita:** 150k yuan (21,000 USD)

Tianjin is a huge port and manufacturing center on the Bohai Gulf, with a significant history due to its key location on the Grand Canal, linking the Yangtze and the Yellow River. Only the fifth largest container port in China [2015], it still shifts over 10 million containers a year, and acts as the shipping gateway to Beijing, only 70 km (40 mi) northwest.

Huge foreign and national investment in manufacturing have made it China's third largest city.

Many of our customers get to Tianjin by international cruise and then transfer to Beijing. We offer Tianjin Port Transfer and Beijing Sightseeing tours.

5.4 Shenzhen — 12.7 million people

Figure 12: Shenzhen.

- **Area:** 1,700 km² (700 sq. mi)
- **Density:** 7,500/km² (19,000 per sq. mi)
- **2018 GDP:** 2.4 trillion yuan (346 billion USD)
- **GDP per capita:** 190k yuan (27,000 USD)

Shenzhen is located in between Guangzhou and Hong Kong. It is a huge manufacturing center that has sprung up overnight. Feeding off the success of its neighbors, it is ranked fourth in China for industrial output, manufacturing higher technology products than Guangzhou in general, and with several of its own successful sunrise companies.

5.5 Guangzhou — 11.6 million people

Figure 13: Guangzhou.

- **Area:** 3,800 km² (1,500 sq. mi)
- **Density:** 3,000/km² (7,800 per sq. mi)
- **2018 GDP:** 2.3 trillion yuan (326 billion USD)
- **GDP per capita:** 200k yuan (28,000 USD)

Guangzhou (Canton) is a mighty manufacturing base, drawing millions from the countryside to work in its factories. Vast quantities of clothing, electronics, plastic goods, and toys are shipped from Guangzhou all over the world. A city that has sprung up recently with China's economic boom, it hosts the biannual China Import and Export Fair or Canton Fair.

It is **the richest city in Mainland China per capita**, still just above Shenzhen. Hong Kong (10th city in the world for GDP per capita) has over twice Guangzhou's 2017 GDP per capita per month (5,000+ USD).

5.6 Chengdu — 10.2 million people

Figure 14: Chengdu.

- **Area:** 1,700 km² (700 sq. mi)
- **Density:** 6,000/km² (15,000 per sq. mi)
- **2018 GDP:** 1.5 trillion yuan (213 billion USD)
- **GDP per capita:** 150k yuan (21,000 USD)

Chengdu is an exception among large Chinese cities. The largest city in mostly mountainous or arid West China, it is a concentration of the population of the fertile Sichuan Basin. The pace of life is the most relaxed of China's large cities.

Though industry does play a part in Chengdu's economy and there has been significant domestic investment, its growth is mainly a result of the tide of urbanization driving the rural population towards the cities in search of a better life. With Chengdu being the lone large city in huge Sichuan Province, the province's over 80 million people gravitate there.

This is a very thirsty market as northern China has little water.

5.7 Chongqing — 8.5 million people

Figure 15: Chongqing.

- **Area:** 1,000 km² (400 sq. mi)
- **Density:** 8,500/km² (22,000 per sq. mi)
- **2018 GDP:** 0.73 trillion yuan (104 billion USD) [based on an urban/rural income ratio of 2.5]
- **GDP per capita:** 80k yuan (12,000 USD)

Chongqing is famous for its fog and mountains, and the Yangtze River cruise. It's the largest of China's four municipalities besides Beijing, Tianjin, and Shanghai, though by contrast its population is mostly rural. Huge infrastructure and industrial investments have made it one of China's 10 largest cities in the last 5 years.

Chongqing is **Mainland China's most densely populated city**. Chongqing Municipality's main city is sucking in its 30-million+ provincial population into an area constricted by mountains and the Yangtze River — so high-rise dominates. (Hong Kong has a greater population density, over twice Chongqing's, once again forced up my mountainous geography and water.)

5.8 Dongguan — 8.3 million people

Figure 16: Dongguan.

- **Area:** 1,600 km² (600 sq. mi)
- **Density:** 5,200/km² (13,000 per sq. mi)
- **2018 GDP:** 0.8 trillion yuan (113 billion USD)
- **GDP per capita:** 100k yuan (14,000 USD)

Dongguan is a little-known but huge manufacturing city, between Guangzhou and Shenzhen, ranked fourth in China for exports. It has also grown phenomenally in the last couple of decades. It employs huge numbers of rural factory workers, producing electronic items and other hardware, like computer peripherals.

It's **the poorest of China's large cities**, dominated by low-wage-earning migrant factory workers. Over a million overseas Chinese and residents of Hong Kong, Macau and Taiwan came from Dongguan. Tourism is virtually unheard-of there, apart from for those returning to their roots.

5.9 Shenyang — 7.9 million people

Figure 17: Shenyang.

- **Area:** 1,500 km² (600 sq. mi)
- **Density:** 5,300/km² (13,000 per sq. mi)
- **2018 GDP:** 0.62 trillion yuan (88 billion USD)
- **GDP per capita:** 80k yuan (11,000 USD)

Shenyang is the capital of the northeastern province of Liaoning, and the largest city in northeast China by urban population. It is another city that has **grown rapidly in the last 5 years** to make the largest 10.

It is known for its status as a leading industrial, commerce, and trade city. It is also home to impressive aspects of Chinese history, being briefly the capital of Manchu-led China and Japanese-occupied China, with its own imperial palace. And it has a Korean minority influence.

5.10 Wuhan — 7.9 million people

Figure 18: Wuhan.

- **Area:** 1,500 km² (600 sq. mi)
- **Density:** 5,300/km² (13,000 per sq. mi)
- **2018 GDP:** 1.5 trillion yuan (213 billion USD)
- **GDP per capita:** 190k yuan (27,000 USD)

Wuhan is an interesting large city, plum in the center of the heavily populated half of China. It once felt less modernized than China's coastal cities, but it is now one of China's main high-tech, education, and financial centers.

It has long been **a transportation hub** because it is in the middle of a long navigable part of the Yangtze River between Shanghai and Chongqing. I've sailed the Yangtze from top to bottom.

This city has a certain historical charm about it and has been a key city since the Warring States Period (481—221 BC) due to its location. The people are unusually friendly compared to those in other cities.

2021 Wuhan became an international interest due to the release of COVID-19 the China Virus.

Although it does not have a costal port, the Yangtze River is huge allowing adequate access.

5.11   A Brief Market History – this decade

In 2008, Chinese consumption of bottled water had already increased by over 11% to an average of 15.5 liters per person. This trend continued and has doubled more than once since, as consumers become more health-conscious, young consumers drive volume growth, with bottled water consumption per person expected to reach 18, then 20 liters by 2010, equivalent to a total market volume of almost 24 billion liters. But these projections were far too low.

The Euromonitor report we have in our budget will update these comparable as of 2020. This report is targeted to three specific markets in China. (Singapore, Hong Kong, Beijing) the next round of market investigations will focus on Tianjin in northern China and huge manufacturing center, Chengdu, Chongqing, Shenzhen, Guangzhou. These markets have more than adequate ports to offload our water.

Challenges
China faces two significant challenges with regards to bottled water industry regulation: 1) the improvement and preservation of water quality and 2) consumer protection from unscrupulous operators of which there are hundreds. Issues surrounding bottled water in China continue without any signs of lessening despite the efforts of Chinese officials. In fact, several wells have been drilled over the past decade to meet market demands, many without official permits and with contamination. This combined with more groundwater pumping in China's coastal aquifers at a rate that exceeds the capacity of rainfall to replenish them in a very short time. This has already caused surface subsidence, saltwater intrusion, and eventually groundwater depletion such as has happened along almost all of China's coastal aquifers.

Bottle vs. Can
The prospect of contamination scares in the bottled water market are very real, especially after the tainted milk scandal highlighting the importance of proper regulation and enforcement. A part of the marketing strategy in the bottled water business is to have **"in-the-can"** a major marketing placement that takes immediate advantage of such scandals whenever they occur. Those who have

used this can package, jump their sales almost immediately and capture greater long-term market share. The problem obviously is you can't 'see' the water you are buying compared to a clear glass or plastic bottle. Our marketing influence will show bottlers how to maximize the visual impact of our 'glacial' water on consumers.

Until the quality of tap water is consistently reliable, not something that will happen in our lifetime if ever given the massive scale of the problem, Chinese consumers will continue to purchase bottled water as a replacement for tap water. They will continue favoring drinking water that can be guaranteed as pure, *natural*, of special quality, and pay the lowest price possible. With the market forecast to grow by an average of 12% per year for at least the next 5 years, this market offers a clear opportunity for new product lines, especially of high quality, top *natural* glacial branding, and even at the top - competitive prices.

Market Studies Needed
Because we find that China is our best and first market, we need to acquire a couple of specifically focused bottled water market studies by independent global product market analysts to better inform us on potential buyers/ partners in several markets within China. As noted earlier, one such study was completed in **May 2009** by **Euromonitor International** and can be purchased for $900. We have been advised that this has been updated. There are others that can also be acquired. At our request, Euromonitor has prepared an updated very targeted product/market analysis for us. In addition, Grail Research has prepared a proposal for a market analysis for our consideration. These proposals will outline not only wholesale prices/costs now in this market, but also potential market partners, other buyers, marketing strategies, locations, costs, time, margins, etc.

We have provided this level of market detail so that you have confidence in our approach to markets. "Know your market" is a guiding mandate in AQUEOUS International, Inc. Something we have learned not just in graduate school, but also by experience watching the state of Alaska fail in this regard as they try to develop resources without any real understanding of market prices or demand over time. Most of these ventures by the state have failed. market'.

Figure 19: "Know Your Market".

5.12 UPDATE:

The richest person is China is now a bottled water tycoon, knocking Alibaba founder Jack Ma from his mantle. Zhong Shanshan founded Nongfu Spring in 1996 from a lake in the Zhejiang province on China's coast just south of Shanghai. The water from this source is from a well below the lake with good mineral content, **but it is not natural Alaskan Glacial Water**. We think this new billionaire will be extremely interested in offering a new even higher end brand of 'fine water'. The Bloomberg Billionaires Index now puts Mr. Shanshan in the top spot with wealth of $58.7bn (£46.2bn). Sep 24, 2020

Figure 20: Chinese Bottled Water.

Figure 21: Chinese Bottled Water.

It's Just Marketing:

One of the best lessons I learned was from the top Scotch 'noser' in Scotland while I was there at the Queen's Hunting Lodge. She was teaching a class on Scotch and had recently resigned from Johnny Walker because she refused to lie about new labeling and marketing strategies by that company. When I asked her what the difference was between several new brands from Johnny Walker she said – nothing, *"It's only marketing"*. In fact, I found out she was correct after looking at some reviews of these 'new' products. The marketing and BRAND set up a consumer to think and feel and even taste that what they are drinking is special – when it may not be. Just look at the above picture of a few of Nongfu Spring water's product line. The presentation is what makes it special, otherwise it's just the same lake water.

This is a key in our BRANDING and one of the main reasons why, when we have market partners, we insist that we be involved in their marketing so that our brand – 'natural Alaskan glacial water' is ALWAYS on the label.

5.13   South Korea:

Preferred Option/due to economic and growing political factors: As a result of several factors and recent events we have looked very carefully at a new angle in reaching the first ten Chinese markets via the west shore of South Korea. Why? The economic and political instability of trade agreements between the United States and China, the deep problem of corruption in businesses in China, the influence of the People's Party of China in almost every business, the growing need for clean water in Chinese municipal water systems starting with Shanghai which consumes well over 90 Million gallons a day is an excellent Muni Mix opportunity, and the recent virus pandemic causing even more political and economic street.

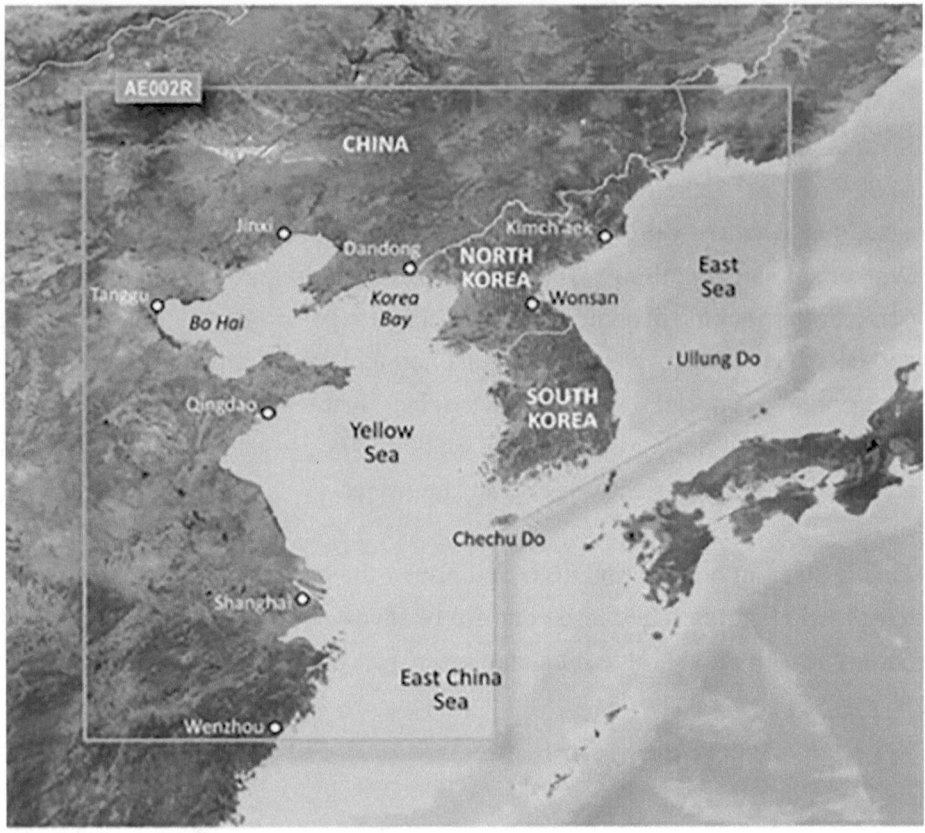

Figure 22: Eastern Markets.

China is only 400 miles across the Yellow Sea from Incheon harbor, S Korea. The waters in this sea are generally predictable allowing conveyance in just a day. When you also consider 35,000 cubic meter WTBs as a towable and/or in-sea storage for municipal mix buyers, this conveyance option enables far more secure and lower cost predictable sailing and thus on-schedule deliveries. No one likes their water to be late.

Korean Drop Off Point

The idea is simple. We deliver pure natural Alaskan glacial water to Incheon harbor in eastern South Korea, have it bottled/packaged there and then shipped in 40-foot containers (just 400 miles) to the ten targeted coastal markets in China we have assessed. We can also look at the injection of WTBs taking water from the VMaxx harbored in Incheon Harbor for delivery to industry and/or municipal water systems in China. WTBs should not have any problems in this crossing and can immediately be integrated into the onshore storage

Figure 23: Chinese Port.

This option will give us far more control and fiscal certainty both in the contracts and in the political stability within South Korea rather than docking in China.

If we find a Chinese market that wants to buy bulk water at the dock in China and we have a 30 year delivery contract, which we understand is NOT enforceable in China, we might think about it, but given all the immediate factors, we think this option presents far more immediate opportunity with security.

NOTE: South Korea is already a major exporter of bottled water into China (see section on competitors). We can consider an option of selling our AK natural glacial water to an existing beverage manufacturer in South Korea provided they use our brand name and then it is exported to China, or we can openly partner with a South Korean company specifically for the China markets. Obviously, there are other options to be discussed. It is our plan to be in South Korea in 2024 to speak at the Korean World Trade Center, delayed in 2020 by the Chinese virus, and initiate these negotiations. We are already working with one Alaskan company that has a relationship with a bottled water company in South Korea and we have many businesses in Alaska owned and operated by former South Koreans.

Figure 24: South Korean Shipyard.

The Hyundai Heavy Industries in South Korea are also the largest builder of the VMaxx design tankers, our preferred vessel for freshwater conveyance for several reasons. (discussed in more detail in the section on conveyance options - page 69)

5.14 Water Transport Bags (WTB) from S Korea to Chinees markets

The application of Water Transport Bag (WTB) such as we developed and utilized in the Mediterranean until the Turkish government changed our 30-year export agreement and attempted to leverage significant cost increases that made our exports to contracted markets uneconomic. Using WTBs will have a particularly good application from South Korea to Chinees markets just 400 miles cross-water. They not only offer conveyance but also in market at port water storage, again lowering the cost to a buyer at start up until they build or provide their own bulk storage as they determine necessary. WTBs can be built to meet extremely specific delivery schedules up to 35,000 cubic meters per lift loaded from VMaxx tankers in the Korean harbor.

It is also possible that we can find a Korean company to manufacture our WTBs for us not just in the Chinese markets but others as well. This will work as an exceptionally good incentive to South Korean investors and others to further develop WTBs for any number of applications in Asia and potentially globally.

This is an active option we are now actively pursuing with a South Korean partner or two.

DISCLOSURE: Chairman Davidge lived and worked in South Korea from 1977 to 1979+. Part of his work involved visiting many towns and seaports from his base in Seoul where he was stationed with 8th Army UN Command Headquarters. He speaks enough Korean to get him in trouble, but it often comes in handy over drinks.

5.15 China Complications

Grey Zones

Over the past decade what is going on in China has caused us concern. Of key interest are their *'grey zones'* and how they leverage debt to 'take' ownership of lands and facilities in other countries. Often for a military strategic purpose but also for an economic purpose to include vegetables and other products. In many poor countries their governance leadership needed money to do something. Build a road, a bridge, a rail system, a port, and a Chinese 'company' offered to help. What these folks did not understand was that the 'company' was in fact an extension of the Chinese Communist Party (CCP) with extremely specific 50-year goals that included this specific area. And so, over the past two plus decades China has gathered lots of natural resources, infrastructure, strategic post locations and state of the art facilities for their longer-term agenda. "War without Restrictions" is a book written by two top Chinees Army Colonels in 1999, following their investigation into America's successful war on portions of Islam, outlines a 'WIN' strategy without any international or moral limitations. If you have not read this book, we strongly recommend it even if you have no interest in China, because they have an interest in you – especially North America.

Pandemic Spread

With the election of Donald Trump as USA President, a man who has long called out for attention to what China has been doing with the consent of 'compromised' politicos, business owners, and others - things have gotten a little more uncertain. The standing order of War without Restrictions continues to be the CCP policy – in fact, some intelligence experts report that the global COVID -19 problem, although likely an error in the Wohlen lab, but when the CCP made the decision to close all of China but ship out anyone who wanted to leave and anyone with the virus to any country in the world – that was a decision consistent with their 50 year policies.

Even with the Bidon election, we are not sure about the political issues between China and the USA. We keep a close eye with many good friends in Congress in Wash, DC.

One of our main concerns is potential vulnerability of water sources in SE Alaska to applications from a Chinese business (well sort of) to take water. Existing state law would allow this, but as the former State Director of Water I am working with our legislature to ensure it is not allowed to any other country. But the details are very much in the small print. We will keep any client or partner advised on any developments here. Our sources will not be in jeopardy, but there are others that may be ripe for picking and we are surveying and assessing them for future interests.

NOTE: The purpose of this short discussion is only to be open and careful and keep prospective investors, clients, and market partners advised. Nothing more. We have excellent sources of intel both in China and in the USA that keep us advised.

5.16 India

Contamination

Tap water in India is not at any reasonable standard that would encourage anyone able to buy an alternative to drink it. Annual studies reveal that even much of their bottled water is contaminated. Tap water is mostly used for bathing, washing, and other domestic uses other than cooking or even animal consumption. This is not likely to change over the next two or more decades simply due to the magnitude of their contamination and the proven economic impossibility to clean this up. The over whelming amount of industrial and human waste being generated is far beyond any capability of control or economic ability to treat/clean it, and the over exploitation of groundwater reserves – especially in coastal aquifers, most of which are already grossly contaminated with pesticides, fecal coli, and saltwater, and more and more pharmaceuticals.

India is generally behind China in its consumer understanding of the quality of contents in bottled waters or beverages but is catching up quickly due to recent discoveries that **most bottled beverage brands in India are contaminated with banned pesticides including DDT and pharma hormones**. According to a recent national study there are close to 200 bottled water brands in India. Nearly 80% of these are local brands **completely reliant on local water** sources with no real government oversight. Most water sources for bottled water are from wells that have long histories of contamination.

Evian, which is not bottled in India, was one of the few brands found in India markets, without pesticides. Additionally, the exponential infusion of pharmaceuticals into the groundwater is a growing and serious health (and for men their potency) concern of the Indian government agencies responsible for ensuring the quality of bottled water. Recent articles pointed out that some of these contaminates cause **serious physical impairments** ranging from damage to the central nervous systems to lung cancer and total infertility.

2019

According to the most recent report, March 22, 2019 by Value-Research, the market is expected to reach INR 403.06 Billion by the end of 2023, from the current value of INR 160 Billion, expanding at a compound annual growth rate of almost 21%. Based on volume, the market is likely to reach 35.53 Billion liters by 2023, expanding at a rate of 18.25% from 2018 to 2023.

> "The genie is out of the bottle. Indeed, the bottled water industry is one of the most thriving sectors in India. The market is growing at a whopping rate of about 55 per cent annually and is expected to cross Rs. 1000-crore mark within the next couple of years. Parle's Bisleri that virtually monopolized the bottled water market, is now vying with Nestle, Coca Cola, PepsiCo, Manikchand, UB and Britannia. According to a national-level study, there are close to 200 bottled water brands in India. Nearly 80 per cent of these are local brands. There are around 150 domestic Indian Bottled Water Brands." (source: Fine waters, India)

We are currently in discussions with Euromonitor for an update (on the Indian markets with more targeted focus on three major cities that we believe would be good opportunities to introduce imported pristine *natural* Alaskan glacial water that will set a new 'brand' standard in this huge market. The water is imported, but the bottling is local, so we MUST ensure our market partner's compliance with our contracts, ensuring it is done at the highest quality. We have also asked Euromonitor to start considering potential manufacturing opportunities while they vet at least three bulk water partners in these three cities (Mumbai, New Delhi, Calcutta, now known as Kolkata).

The bottled water industry in India, predominately in three cities, is estimated at about **$203M USD** and is **growing at a rate of 40% - 55% annually** depending on the study you read.

2019 The Bottled Waters of India

Aava	Kenbar
Aion	Kingfisher
Alankar Phoenix Aqua	Kinley
Alfa & Omega	Maqua-2000
Alfa Blue	MInerwa
Aquafina	Nakshatra
Atlas Premium	Natural Spring - Natural Mineral Water
Avian	Nestlé Pure Life
AxyZen	Palmspring Mineral Water
Bailley	PingX
Basil	Pristine
Bibo	Pure Lifekare
Bisleri	Quench Mineral Water
Cool Valley	Quinse
Dew Drops	Raindrops
Dislaren	Rivano
Edlin	River Bank
Fit-n-Fine	Sahil
Golden Eagle	SB JAIN
Golden Valley	Silver Springs Aqua
Haywards 5000	Sindhu Safe Water
Hello	Snow Pure
Himalayan	Spaa Aqua
Hind Neer	Sprint Up
Jairu Naturelle	Universal-Aqua
Just Born Spring Drops	Vcare
Karni	Yelgris

Figure 25: Bottled Water of India.

Look at these brand names. Is there anyone among them that can compete with 'pristine *natural* Alaskan glacial water' and its emotional imagery?

Indian Brands

Ramesh Chauhan's, **Bisleri**, the bottled market leader, ranks, in contamination, 15th amongst 17 water brands studied in Delhi. It was found to have a total pesticide content **79 times more** than the prescribed limits and seventh amongst the 13 brands studies in Mumbai.

Even Coca-Cola's Kinley brand has been ranked 8th in contamination in Delhi with 14.6 times more than the prescribed limits of pesticides, while it is on the 5th spot in the Mumbai list of contaminated bottled waters. Coca-Cola admitted that pesticide residue was a national problem but insisted that they are maintaining high quality standards in all their plants. They are not.

Other brands listed in this critical study included No. 1 McDowell and Kingfisher from the UB stables, Pepsi's Aquafina, Nestle India's Pure Life and DS Foods' mineral water Catch. All the No. 1 McDowell products were found to comply with the standards.

In Delhi, most water companies depend on well water that are in industrial or agricultural areas giving context to the levels of pesticide contaminates. In Mumbai, however, the companies use water supplied by municipal sources that **only clean between 20 to 80 percent of the pesticides out of their source water**. They, like the rest of the world, have no capability to remove pharmaceuticals from the groundwater.

Generally, the bottled water market in India is lucrative and growing fast with excellent long-term prospects. **Finding new clean sources for their feed stock is critical to maintaining any public confidence in their products.** The first company that is located within a port complex with the capacity to import *natural* Alaskan glacial water as their source/feed stock has an extraordinary opportunity to capture huge market share even at the mid to top of the market. This is our opportunity and the opportunity of our market partner.

An important summary of the Indian Bottled Water Industry is provided below. Written by Chandra Bhushan, the Associate Director, Centre for Science and Environment, New Delhi.

At the fourth World Water Forum held in Mexico City in March 2006, the 120-nation assembly could not reach a consensus on declaring the right to safe and clean drinking water as a human right. Millions of people the world over do not have access to a potable water supply. But this is good for the bottled-water industry, which is cashing in on the need for clean drinking water and the ability of the urban elite to pay an exorbitant price for this extremely basic human need.

The More we pollute, the more we need bottled water

The fortunes of this more-than-$100-billion global industry are related to the human apathy towards the environment - the more we pollute our waterbodies, the better the sales of bottled water. It is estimated that the global consumption of bottled water is nearing 200 billion liters - sufficient to satisfy the daily drinking water need of one-fourth of the Indian population or about 4.5 per cent of the global population.

In India, the per capita bottled water consumption is still quite low - less than five liters a year as compared to the global average of 24 liters. However, the total annual bottled water consumption has risen rapidly in recent times - it has tripled between 1999 and 2004 - from about 1.5 billion liters to five billion liters. These are boom times for the Indian bottled water industry - more so because the economics are sound, the bottom line is fat, and the Indian government hardly cares for what happens to the nation's water resources.

India is the tenth largest bottled water consumer in the world. In 2002, the industry had an estimated turnover of Rs.10 billion (Rs.1,000 crores). Today it is one of India's fastest growing industrial sectors. Between 1999 and 2004, the Indian bottled water market grew at a compound annual growth rate (CAGR) of 25 per cent - the highest in the world.

Growth in the bottled water demand in India

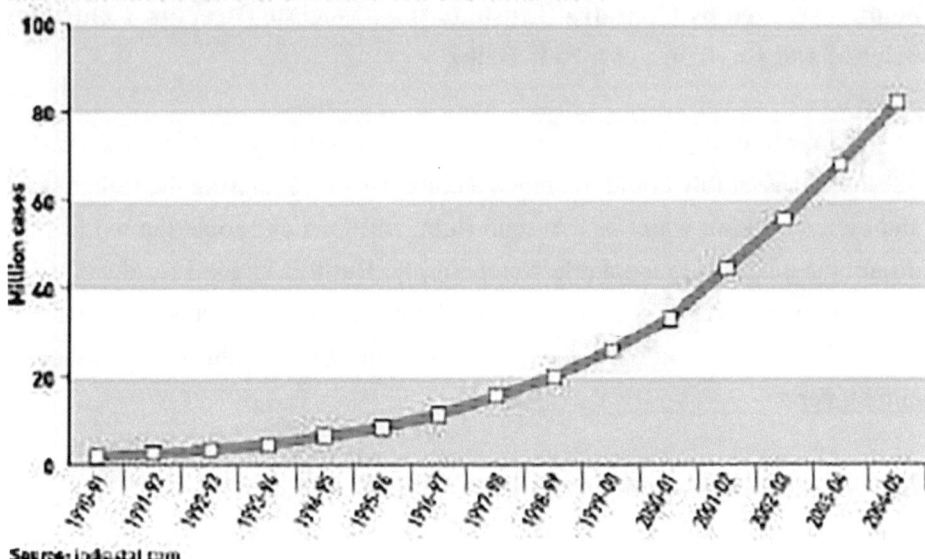

Figure 26: Growth in Bottled Water Demand in India.

With over a thousand bottled water producers, the Indian bottled water industry is big by even international standards. There are more than 200 brands, nearly 80 per cent of which are local. Most of the small-scale producers sell non-branded products and serve small markets. In fact, making bottled water is today a cottage industry in the country. Leave alone the metros, where a bottled-water manufacturer can be found even in a one-room shop, in every medium and small city and even some prosperous rural areas there are bottled water manufacturers.

Despite the large number of small producers, this industry is dominated by the big players - Parle Bisleri, Coca-Cola, PepsiCo, Parle Agro, Mohan Meakins, SKN Breweries and so on. Parle was the first major Indian company to enter the bottled water market in the country when it introduced Bisleri in India 30+ years ago.

The rise of the Indian bottled water industry began with the economic liberalization process in 1991. The market was virtually stagnant until 1991, when the demand for bottled water was less than two million cases a year. However, since 1991-1992 it has not looked back, and the demand in 2004-05 was a staggering 82 million cases.

Bottled water is sold in a variety of packages: pouches and glasses, 330 ml bottles, 500 ml bottles, one-liter bottles and even 20- to 50-litre bulk water packs. The formal bottled water business in India can be divided broadly into three segments in terms of cost: premium *natura*l mineral water, *natural* mineral water and packaged drinking water.

Premium *natural* mineral water includes brands such as Evian, San Pellegrino, and Perrier, which are imported and priced between Rs.80 and Rs.110 a liter. Natural mineral water, with brands such as Himalayan and Catch, are priced around Rs.20 a liter. Packaged drinking water, which is nothing but treated water, is the biggest segment and includes brands such as Parle Bisleri, Coca-Cola's Kinley and PepsiCo's Aquafina. They are priced in the range of Rs.10-12 a liter.

Attracted by the huge potential that India's vast middle class offers, multinational players such as Coca-Cola and PepsiCo have been trying for the past decade to capture the Indian bottled water market.

Today they have captured a significant portion of it. However, Parle Bisleri continues to hold 40 per cent of the market share. Kinley and Aquafina are fast catching up, with Kinley holding 20-25 per cent of the market and Aquafina approximately 10 per cent. The rest, including the smaller players, have 20-25 per cent of the market share.

Consumption of bottled water in India is, as is expected, linked to the level of prosperity in the different regions. The western region accounts for 40 per cent of the market and the eastern region just 10. However, the bottling plants are concentrated in the southern region - of the approximately 1,200 bottling water plants in India, 600 are in Tamil Nadu. This is a major problem because southern India, especially Tamil Nadu, is water starved.

Most of the bottling plants - whether they produce bottled water or soft drinks - are dependent on groundwater. They create huge water stress in the areas where they operate because groundwater is also the main source - in most places the only source - of drinking water in India. This has created huge conflict between the community and the bottling plants.

Private companies in India can siphon out, exhaust and export groundwater free because the groundwater law in the country is archaic and not in tune with the realities of modern capitalist societies.

The existing law says that "the person who owns the land owns the groundwater beneath". This means that, theoretically, a person can buy one square meter of land and take all the groundwater of the surrounding areas and the law of land cannot object to it. This law is the core of the conflict between the community and the companies and the major reason for making the business of bottled water in the country highly lucrative.

Coca-Cola

Take for instance the case of Coca-Cola's bottling plant in drought-prone Kala Dera near Jaipur. Coca-Cola gets its water free except for a tiny excess (for discharging the wastewater) it pays to the State Pollution Control Board - a little over Rs.5,000 a year during 2000-02 and Rs.24,246 in 2003. It extracts half a million liters of water every day - at a cost of 14 paise per 1,000 liters. So, a Rs.10 per liter Kinley water has a raw material cost of just 0.02-0.03 paise. (It takes about two to three liters of groundwater to make one liter of bottled water.)

Costs

However, water is not that cheap in the United States, home to Coca-Cola and PepsiCo. The average cost of industrial water in the U.S. was Rs.21 per 1,000 liters in the late 1990s. It was Rs.90/1,000 liter in the United Kingdom and Rs.76/1,000 liters in Canada.

Treatment and purification accounts for the next major cost. Even with the *state-of-the-art multistage treatment systems with reverse osmosis and membranes, the cost of treatment is a maximum of 25 paise a liter (Rs.0.25/liter). Therefore, the cost of producing 1 liter of packaged drinking water in India, without* including the labor cost, is just Rs.0.25. In a nutshell, in manufacturing bottled water, the major costs are not in the production of treated and purified water but in the packaging and marketing of it.

The cost of a bottle, along with the cap and the carton, is the single biggest cost - between Rs.2.50 and Rs.3.75 for a one-liter bottle. For water sold in big plastic jars (20-50 liters), which are also reused, or in pouches, this cost is much lower. It is precisely owing to this that companies sell water at even Re.1 a liter in a 20-50-liter jar and still make profits. Labor and establishment and marketing costs are highly variable and depend on the location and size of companies. Informal discussions with industry members reveal that the gross profit of this industry can be as much as between 25 and 50 per cent.

The reason that companies do not have to bear the cost of the main raw material - water - has made this industry highly profitable. But the real cost of the industry is huge. But this will soon change as local and national governance has realized an income opportunity and it is highly likely that the 'free water' history will soon be over.

Fast Depleting Ground Water
They have not paid close attention to their fast-depleting groundwater, which is incalculable because they do not meter their use, neither do they consider the cost of disposal of plastic bottles and pouches. These are hidden costs that Indian society is not yet willing to address, but with a new income stream initiated by governance to charge for taking 'their' water, this will change. But the most immediate concern is the depletion of India's coastal aquifers and their predictable saltwater intrusion ending that massive source.

Again, it will be our marketing that steers demand. The notion of being able to choose a pristine *natural* Alaskan glacial water as the top brand in India will move a measurable part of this market up into mid to high-end 'fine' waters. Our strategy to entering the market with stories of bottled water contamination as 'news' both on general media and social media tools will prepare this market for our *natural* Alaskan glacial water. The development of this marketing strategy is already underway as we believe it generic to our entry into almost any market with the history of either ground or surface water contamination or the contamination of bottled water and beverages manufactured in country using local water.

Groundwater in India remains the cleanest and cheapest source for all, but they have over-extracted and polluted it with the regular range of contaminants, fecal coli, agra-chemicals, pharmaceuticals - mainly hormones, and heavy industrial wastes. They may try to recharge and revive their groundwater sources but with little success as every effort to do this around the world has failed. In addition, their archaic laws dealing with water will need to change and given the political corruption we don't think this will be soon.

Vicious Sewage Cycle
Like China, their surface water bodies are in a deplorable condition. Water borne diseases are everywhere. They dump their sewage, dead animals, agricultural and industrial waste in rivers and ponds, and then try to clean them in massive, centralized treatment plants to supply mostly demineralized water to urban households - to be discharged again as wastewater into the same water body that is there principle intake source. This vicious cycle does not appear of even interest so far. The cost of cleaning up dirty water is just too great for this society to bear. Bottled water and domestic treatment systems are far cheaper as well as fill-and-forget solutions for over 40 per cent of the population, but in doing so they have not left any solution for the 60 per cent of the poor. One major national cholera outbreak (the most common water related killer) can change this quickly.

India will be a good market for AQUEOUS International, Inc. provided we find competent and trustworthy market partners. Government corruption is systemic, again why we require a market partner so that we do not have to deal with any of these concerns.

AQUEOUS International, Inc. has an incredibly good longtime friend and former associate in the United States who is deeply connected to Indian politicos with a father who served as the ambassador of India to Australia. He will be an important asset in our introduction into this market and we are, and he is, confident this will be a successful market introduction and delivery contracts signed. We will secure his talents and contracts when we are ready.

5.17 Middle East

Solid Partnerships

The business community in the Middle East has been very aware for years of the value of high-end bottled water and beverages. This is represented on the conference tables in every business office we have visited in many counties in this region – and they are many. The quality of the bottled water on the conference table is a statement about the sophistication of that company. Our work with several companies in Saudi Arabia, Jordan, Libya, Tunisia, and Dubai gives us a high degree of confidence that solid partnerships can be built with well-established beverage manufacturers throughout this region. Bottled water is a staple in almost all cities, homes, and businesses across the Middle East. You will see it in the saddle bags of camels as they leave for the desert. But there are other opportunities here as well other than just bottled water.

The increasing global consumer demand for a healthy option to sugary soft drinks, the heightened awareness of the need to stay hydrated in very warm climates (hell it was 110F when I got up in the hotel one morning) and the greater accessibility of bottled water in emerging economies have driven strong growth in bottled water globally. The global growth by bottled water in recent years has been supported by strong growth in bulk formats, such as water for coolers, and the growing familiarization with small pack 'everyday hydration' in all corners of the globe. Access to safe water remains a challenge in many parts of the developing world. In the Middle East bottled water is fundamentally relied upon through a variety of pack formats, sizes, and distribution channels and almost all are imported.

Now, the growing awareness of the health impacts of demineralized waters (desalination) on pregnant women and your children, is turning consumers more and more toward *natural* water for drinking. But recent discoveries, now proven globally, show that the brine discharge of most RO desal plants is greater by 50% than what the industry said it would be. This is having a significant adverse impact on the desal community due to escalating costs, reduced life of membranes which are costly, and the growing local and international concern that this much brine is causing significant harm to sea life.

Many of the water bottling plants in the Middle East have won international awards for their production efficiencies and innovative marketing. In this part of the globe the bottled water community is always well represented at every international bottled water meeting.

Conveyance Costs

Our problem is getting water from Alaska to the Middle East at a cost/price at point of sale that works. We do not think this is yet economically viable from Alaska, but we do have alternative water sources in the Mediterranean and Atlantic that are significantly closer, available, and of comparative quality although not with the *natural* Alaskan glacial brand. Although we are open to bulk imports to this region, we are still looking at costs relative to Alaskan water verse source alternatives based closer to these markets. Even shipping across the arctic when possible.

We are confident that with other sources in the Atlantic there may be an alternative source that works. We do think that countries dependent on desalinization should seriously consider 'mixing' as a water quality step in the right direction. The injection of only a small percentage of Alaskan natural water into the produced waters of a desalination plant can change the health impacts of desal significantly. But we have sources, that unfortunately we cannot own because of their national origins, but can secure via an export permit (which is vulnerable to political change) that could seriously enhance the quality of the drinking water from desalting plants.

Most attractive is a source that has contacted AQUEOUS International, Inc. as prospective bulk export source in east Canada. Others in the Mediterranean have also expressed strong interest in developing their waters for bulk exports.

When we find a capable market partner in the Middle East, we will introduce these options, under confidence, as potential options.

Additionally, we continue to explore freshwater discharge subsea vents and a harvest technology we believe can be safely used to fill an appropriately sized tanker. A VMaxx can transverse the Suez Canal, but other scale tankers can move a great amount of very high-quality water from the Atlantic and Mediterranean regions to the Middle East cost effectively. It is important to remember that in most cases these subsea vents are not within the legal reach of a neighboring country, and many have water that comes from huge limestone systems providing a *natural* water of exceptional quality. These vents are not that hard to find if you ask local fishermen because that is often where they like to fish.

We believe that the development of subsea freshwater vents offers significant and secure new market opportunities to AQUEOUS International, Inc. and are anxious to move into this part of our future.

5.18   California/Nevada et al

AQUEOUS International, Inc. has for some time been working with various interests, large and small, in California and in Las Vegas where we would use a 'Wheeling Agreement' or in-stream trade in the Colorado River that would provide water to Las Vegas and its surrounds. The basic idea is the more water CA brings into San Diego by tanker with us, the more Colorado River water could be left available to Vegas. The nice turn is that LV pays for the water coming into Sand Diego as a trade example.

The obstacle to this and any coastal American market is the **Jones Act**, passed by Congress in 1917 to protect what had been built mainly on the east coast to build ships for war. It's time for an amendment.

5.18.1   The Jones Act is a federal law

It regulates maritime commerce in the United States. The act requires goods shipped between U.S. ports to be transported on ships that are built, owned, and operated by United States citizens or permanent residents. There are no new tankers being built in America for several reasons, only one being cost, operational labor requirements, very strict compliance mainly dealing with oil conveyance, and more NADA!

Jones Act Exemptions

We have suggested a congressional amendment to the Jones Act that exempts the shipment of freshwater from American sites to American markets. The immediate desperate need for more water in southern California is the principal target, but there are others on the Pacific who have expressed serious interest. Our conveyance costs would be far less then reaching Asian markets, so this is extremely attractive and one of the reasons we want 20 sites 'under paper' allowing us to meet new American market demands.

We are now working with a growing number of exceptionally large businesses in CA that want to get this exemption passed ASAP. Our agent in San Francisco is working on this daily. The problem we face is the recall of the governor and all the political problems that he is causing. We will continue to try and move this forward though friends in the US House and Senate.

This is not an easy challenge because of the general commitment to support American shipping businesses, but when you look at the cost differences between a VMaxx built in S. Korea and what one might cost in America we are looking at no less than a 30% difference and likely another 12 months. But when you then add all the compliance requirements imposed not just by federal but also state agencies, the cost just keeps climbing.

We will continue to support this effort and hope that in 2024 we will see some progress in the California Assembly to call on Congress to allow this exemption. Unsure of what the next administration will do in response, but we are already in focus with a number of candidates we know well.

5.19 The Global Market Perspective

In our continued relationship with Euromonitor International, who publish an annual report on the bottled water industry and how it is doing in over 160 countries/markets, we are discussing a new contract that will be far more focused on specific <u>submarkets</u> on a short list in high demand accelerative growth areas.

Booming

Bottled water represents the fastest growing segment of the global beverage market. World bottled water markets were expected to reach $65.9 billion by 2012, stimulated by rising population, consumer spending patterns, lifestyle trends, and growing levels of health consciousness, among others. The market was expected to grow rapidly in the coming years due to growing consumer concerns about fitness, water quality, and health. [Resource: **Global Industry Analysts, Inc**.] It grew faster, much faster than expected but then came COVID.

San Jose, CA (PRWEB) November 10, 2008 -- In the refreshing drinks market, bottled water was the most vibrant segment with market share of almost 40%. Several bottled water companies are looking to enter developing markets with immense growth potential such as Asia and Middle East. China is one such market where sales of bottled water are rapidly increasing. Companies are trying to capitalize on weather changes such as summer and increase in health consciousness amongst consumers. The bottled water industry is expected to witness greater participation from soft drink and dairy processing companies, fostered by economies offered by existing international distribution network, and processing equipment.

As stated by the recent report published by Global Industry Analysts Inc., the global bottled water market is dominated by Europe and United States, which together account for about 55% of the market value estimated in the year 2008. Bottled water will continue to fare well in the global beverage marketplace as a healthy alternative to carbonated soft drinks. Gains in bottled water market will also come from flavored varieties and convenient packaged formats, especially single-serve packs.

These projections were low. As soon as possible we will acquire several recent market assessments to provide us with essential intel on growth projections in markets and allow us to modify our market targets if need be. However, China as the world's largest bottled water consumer will likely continue to be our first market and it is our plan with Euromonitor to focus specifically on our ten 'first' markets. This work is underway.

5.20 Other Strategic Market Reports

We have identified many recent reports on various aspects of the water and bottled water industries in China and are in pursuit of a number of those that are available free. Most require we purchase them, at costs of up to and including $900 each. We have found them worth the investment.

Although we like Euromonitor and continue to work with them, as we pointed out earlier, we are also in discussions with Grail Research a company we have researched and have secured a preliminary proposal at $45,000 for additional market research if we find it necessary. We have additional work to do with Grail before we would authorize this work with them and will wait until we have the updated and purposefully focused Euromonitor reports, but we are convinced that the more we know the better our strategy in selecting buyers/partners, structuring our marketing, in China and the India markets. At this point, we will start with Euromonitor as we have found them deeper, broader, and more responsive to the study of submarkets in massive cities that can turn to our advantage.

5.21 Bulk Water to the United States

Market Report Overview-from GWI

Again, the legal barriers to any bulk market in the USA is the Jones Act, a federal law that regulates maritime commerce in the United States enacted in 1917 following WWI to help protect the growing US shipping industry, mainly on the Atlantic coast, and their union. The Jones Act requires goods and people shipped between U.S. ports to be transported on ships that are designed, built, owned, and operated by United States citizens or permanent residents. The cost variance of a new tanker that would meet these restrictions is now over 30% higher than one built, as most in the world are today, in South Korea. This cost variable completely changes the economic viability for bulk exports from Alaska. We have talked with San Diego, L.A. harbor, and Santa Barbara about importing bulk water as a municipal mix option along with a growing number of wholesale and retail businesses in this market.

5.21.1 Water Market USA

Aging Infrastructure

Spending on water and wastewater in the USA is the highest it has been in a decade plus. It is anticipated that a total of $650 billion will need to be spent over the next 20 years to tackle aging infrastructure, to meet regulatory and water quality demands and to adapt to a changing climate, making this one of the most significant markets in the global water sector.

Covid-19

In the short term, the COVID-19 crisis had significant implications on the shape of the US water market, but it has also created opportunities for private sector businesses who can support utilities with constrained budgets in recovering lost revenues, reducing costs and in finding new sources of revenue. This report is an essential study of the implications of the pandemic on the US water sector, and it also provides the most comprehensive long-range analysis of the US market on a national and state-by-state level.

Water Market USA examines the state of the US water market as it is now, its key drivers of change, and how it will be shaped by COVID-19 over the coming years. As the impact of COVID-19 plays out across the US Water sector, businesses need access to comprehensive market intelligence, reliable forecasts and detailed analysis on drivers, restraints, and challenges to ensure they can identify and capitalize on any new opportunities.

To minimize the pandemic's effects on the sector, utilities have shifted their practices to encompass social distancing measures, creating new opportunities in automation and digital control. The sector is also keenly exploring means of reducing costs, recovering lost revenue, and generating new income, which is explored in the report's detailed assessment of the COVID-19 impact. Strong fundamental drivers such as PFAS, aging infrastructure, climate change concerns and water scarcity in the Southwest will continue to drive growth in the sector beyond the pandemic, creating opportunities for those who can get involved at the earliest stage.

Featuring a complete assessment of COVID-19's impact on the US water market, key trends, high-level industrial coverage and state-by-state forecasts and opportunities for the 20 biggest states, Water Market USA provides the detailed analysis essential to gaining a solid understanding of the market dynamics at play and to future-proofing your business. This report is a key resource for those active in single or multiple sectors in the US and for foreign and local businesses alike.

5.22 Bottled Water: A Global Strategic Business Report

The global bottled water market did not end 2017 without surprises. After years of strong growth in the USA, bottled water sales surpassed carbonated soft drinks to become the largest beverage category by volume in 2016. And the numbers did not stay idle. From 2014 to 2017 due to increasing concern regarding various health problems caused by consumption of contaminated water, the global bottled market **grew to over $200 billion following 9% yearly growth**, according to the report on the bottled water market from The Business Research Company. In addition to health concerns, rising disposable income also let people in the Asia Pacific region influence the growth of the market significantly.

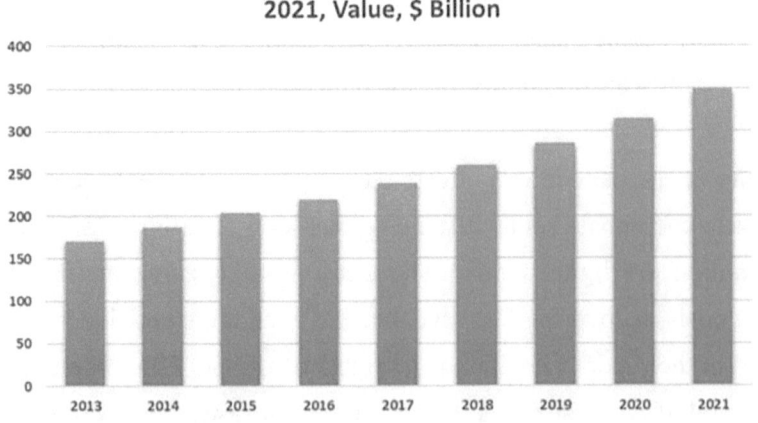

Source: The Business Research Company

Figure 27: Global Bottled Water Market 2013-2021.

By volume, the market grew by 100 billion liters. The predictions for the bottled water market are above previous expectations. By value, **the market is going to reach around $350 billion by 2021**, following 10% year-on-year growth. The volume's growth will be a bit lower, but not by a significant number, at 9.3. Global per capita average bottled water expenditure will also be in a growth stage to 2021, raising to $45.3 by 2021 from $32.3 per head in 2017.

2017 Consumption

The consumption of bottled water in 2017 was the highest in the Asia Pacific region, accounting for 42% of global consumption. The Asia Pacific region is the host for two of the most populated countries where poor public infrastructure is common and access to clean drinking water is limited, all of which drove the market to such numbers. These conditions raise consumers' concerns and prompt them to seek out clean *natural* drinking water to maintain a healthy life.

Following the health awareness trend, bottled water consumers in the Asia Pacific region are also increasingly adopting new products that also provide health benefits, such as functional water, which has added value in the form of minerals, oxygen, and vitamins. Most of these 'are just marketing' and provide no real health value. But since consumers prefer to try new tastes and follow the best of them, the availability of a range of flavors and options in functional water will boost market growth.

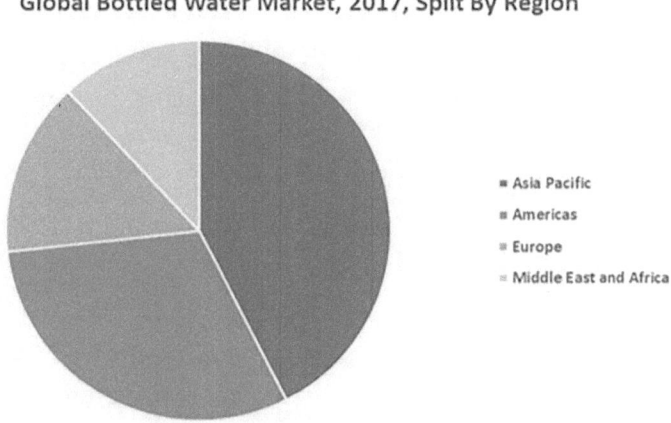

Source: The Business Research Company

Figure 28: Global Bottled Water Market Split by Region (2017).

The second region ranked in terms of bottled water volume was the American continent. High average prices per liter for drinking water made the region rank the first in terms of value.

The global bottled water market is highly fragmented with many local players present in most countries. To stand out among numerous competitors, companies have started changing the look and feel of packaging bottles to get more traction in the market. The aim of producers is to establish a brand image and differentiation in the bottled water market.

Innovative packaging includes re-sealable and recyclable cans or paper packaging material, transparent sophisticated bottles, and inks and coatings applied on a can to enhance customer experience through unique visual effects along with increasing shelf appeal. The leading bottled water companies in the global market that have a multi-country presence are DANONE, Nestle, Coca-Cola, and Pepsi-Co.

AQUEOUS International, Inc. has spent some time and invested in the 'marketing' work we will do with our introduction into the China and India markets with pristine *natural* Alaskan glacial water. One long participating partner has been in the bottled water business for many years using exceptional water from one of the oldest (32,000 years) glaciers in Alaska. Together, our market research will be offered to our market patterners so that they can move quickly in putting our water on shelves with an image that will instantly draw buyers. Additionally, our presentations at regional and global water shows/conferences will be the point of interest and we anticipate initiating many new contracts for bulk delivery.

- We are awaiting new global water updates and will include them upon receipt or in our next update.

2016
Global distribution of bottled water based solely on volume consumed reveals that North America consumes over 30%, Europe consumes over 30%, Asia consumes over 26%, and the rest of the world consumes about 13%. (but remember China has now taken over first place based on total consumption)

Sparkling Water

The sparkling water market is dominated by Europe, which accounts for more than 75% of the world market estimated in 2008. Western Europe comprises some of the world's largest per capita bottled water consumer markets, which include Germany, France, Spain, and Italy. The largest proportion of bottled water consumers can be found in Germany as 88.2% of its adult population drinks bottled water. The non-sparkling water market constitutes the bulk of bottled water shipments, accounting for between 80-85% of the market in both volume and value terms. Non-sparkling bottled water is also expected to offer the highest growth opportunity, outgrowing the sparkling bottled water market by three to four times. Market for non-sparkling water in United States is projected to reach US$15.1 billion by 2012.

Good Water and Health

The major challenge for most companies is product innovation and differentiation as water is still just water, but that is changing as folks learn about what water is in its context from source and either filtration or desal. The growing international concern with the impact of less than good water on human health especially pregnant women, the unborn child, the child for its first two years, the growing concern with male impotence as sperm counts fall – all these marketing points that we have embraced, enable immediate impact on well-established markets.

But AQUEOUS International, Inc. is not just water but the provider of a <u>natural</u> (not filtered) glacial Alaskan water with all of its organic minerals, that will easily attract buyers.

Containers are an important part of the bottled water market, as they constitute nearly 47% of the cost at retail. Even its shelf look, weight and price of the product are as significant as the water itself. For instance, Colorado-based Biota sells bottled water in biodegradable bottles that are produced using corn to attract consumers who are ecologically conscious. Some companies try to attract children by selling bottled water in attractive bottles with vivid designs and colors.

Industry Consolidation

The bottled water industry is in a continuous state of flux. Entry barriers are low and decreasing by the day. Large companies are consolidating by acquiring local/small/niche players to increase capacity as well as product portfolio. Consolidation in the industry is expected to accelerate, resulting in more concentration, with the leading players strengthening their global presence. The bottled water industry in the Asia-Pacific region is expected to experience a major boost, as more companies venture into the market to capitalize on the latent demand potential.

But we offer a product that is not yet in the market, regardless of what some international producers may claim. We know their sources and can, if necessary, review them in our marketing – yes including Coca-Cola, PepsiCo, Icelandic, and more.

Product, marketing/promotion, packaging, distribution, price, and technology would be the key factors in garnering market share in the industry. With the ever-increasing presence of soft drink majors, such as, Coca-Cola, and PepsiCo, and established leaders Nestle, and Danone, the market is expected to become a cauldron of action in the future.

Key Players

Key players dominating the global Bottled Water market include Aqua Gold International Inc, Boreal Water Collection Inc, China Water & Drinks Inc, Groupe Danone, Danone Naya Waters Inc, Isbre Holding Corp, Nestle SA, PepsiCo, Quilmes Industrials Sa-Adr, Saint Elie, San Miguel Corporation, and the Coca-Cola Company.

The report titled "Bottled Water: A Global Strategic Business Report" published by Global Industry Analysts, Inc., provides a comprehensive review of market overview, key market trends, product overview, product introductions/innovations, and recent industry activity. The report analyzes market data and analytics in value and volume sales for regions such as the United States, Canada, Japan, Europe, Asia-Pacific (excluding Japan), Latin America and Rest of World. The study also analyzes the Bottled Water market by the following product segments - sparkling water, and non-sparkling water.

At this point there is no global entity willing or capable of approaching the opportunities worldwide we have and continue to explore of importing high-end natural glacial freshwater stocks from ALASKA. In most cases water markets, including bottled markets, are served locally with marginal technologies and water sources that are generally not secure or of any remarkable quality. In fact, annual studies continue to show significant contamination even in imported brands.

We estimated some time back that the world would undergo significant change in 2015 and beyond as most people and their economies on this planet were without sufficient water to prosper. The growth in the bottled water markets grew faster indicating that the maturation of drinking water markets and consumers and their growing awareness of the health value of consuming better water advances even with fake news.

Three Year Maturation for AQUEOUS
It will take at least three years for this new company to position itself to be **THE** global natural glacial water company and more, either for bottlers, manufactured products, or municipal waters with the capability and vision to capture this moment in time with freshwater delivery contracts that will run beyond our lifetimes - generating significant incomes, and generational good paying jobs. It will not just be marketing, but developing markets and regional and national, and international media interest in what we are doing. We will do that, especially at the water shows and meetings that often 'present' new products for global markets.

We have an exceptional media professional who has worked for us off and on as we needed her. She has spent time in Alaska and developed very good material for all media. She is now on retainment and we hope to have her back full time asap.

Think for a moment, if you had controlled the water sources and their conveyance to Los Angeles that enabled it to become one of the largest economies on this earth, think of the economic implications. This is the kind of opportunity we are looking at in dozens of major global markets. Water is not only necessary for life; **it is essential for economies**. If you shut off the water to Los Angeles, the economies there would collapse, and it would return to a desert. This is exactly what is and will happen in many areas of the world in the next two decades, unless, what we call 'new water' presents.

Their choices are to import the water, desalt/demineralize the sea at extraordinary economic and environmental cost per cubic meter, recycle wastewater (toilet to tap) which has now proven seriously problematic even in its use in agriculture, rationing water, or pack up and leave. Now that is a market that is ready for a company with the capability and vision to step forward and offer what they need to economically survive and prosper. Keep in mind that good freshwater is like a drug; once you have it and are dependent on it, you just want more.

By starting this venture with extraordinarily high-quality pristine *natural* Alaskan glacial water for mid to high-end beverages in these massive regional markets, we immediately will establish this venture with global credibility. Moving then to other means of providing freshwater builds on that credibility in markets where we will have **proven profitable in market partners.**

Another World View - By John Karolefski [15-Apr-2002]

Although this article is before the downturn in the global economy and some changes in regional consumption in the U.S. markets, I still think it provides a good context for what we are proposing. **_It's all about perception i.e. marketing._**

On the surface, water does not seem extremely exciting but bottle and brand it and the stakes change considerably. Consider that a liter of water costs more than a liter of gasoline, and that a typical Western European drinks 17 to 18 cases of bottled water each year, while the average American drinks only ten cases. Now consider that the US market is projected to double in two years and suddenly the business of water looks a lot less dry.

Competitors

At one end of the spectrum of competitors are brands with European roots led by Evian, the world's best-selling water. At the other end are niche and regional brands with limited distribution, such as Minnehaha, a 106-year-old spring water from Hunting Valley, Ohio. Altogether, hundreds of bottled waters will engage in a battle for brand supremacy, or more accurately, brand survival.

Two companies dominate the world market for branded bottled water: Groupe Danone, based in Paris, and Nestle, S.A., based in Vaud, Switzerland. The former markets Evian, Volvic, and others, including Dannon spring water. Nestle markets such majors as Perrier, Vittel, San Pellegrino, Poland Spring and Deer Park.

The US subsidiaries of these global giants are jockeying for shelf space with private labels, regional and niche products, and with two formidable soft drink giants: Coca-Cola and PepsiCo, marketing their own brands of water.

Bottled water falls into two broad categories: non-carbonated (distilled, filtered and spring) and carbonated (both naturally occurring and mechanically added). The war will largely be fought among the non-carbonated brands, which account for most of the volume.

Because it is almost always about the marketing, we add this short item for perspective. What is driving the market for bottled water in the US?

"There is a public perception that municipal water is not as safe as it could be," explains Walt Boyes, principal of Spitzer and Boyes, consultants in water treatment and distribution. "In some cases, that's true. There are many places in the US where the drinking water coming out of the tap is barely drinkable. In other places, that is not true at all. New York City's water is every bit as good as bottled Evian."

European Water Status

That has hardly been the case in Western Europe where bottled drinking water has been standard for years. Even until the early 1970s, there were large areas in Europe – especially Italy, Spain, and Portugal – where it simply was not safe to drink water from the tap, according to Boyes. That situation gave rise to the great European brands of sparking mineral water such as Perrier of France, San Pellegrino of Italy, and Gerolsteiner of Germany. All of these can be found on the shelves of supermarkets with average assortments of water.

The US began disinfecting its municipal water in the early 1900s, making tap water safe to drink. But interest in bottled water began growing in the last few decades, experts say, because of several factors – a decline in drinking water quality in some regions, public perception, healthy lifestyles, the need for proper hydration after exercise, and effective marketing campaigns by makers of successful European brands. Perrier is credited with igniting the bottled water market in the US in the late 1970s.

The reason why there are so many brands of non-carbonated water is that getting into the water business is inexpensive, according to Boyes. A bottling machine costs $100,000.

"What keeps stores making their own brands is the profit," he says. "If a store can make its own water in its distribution center for 10 cents a bottle and sell it for US$ 1.25 a bottle, they'll make more money than if they're selling [name brand] water at a three percent margin."

Two Water Sources

According to the International Bottled Water Association, there are two types of sources from which bottled water can be drawn. One is natural sources such as spring and wells.

The other is approved potable municipal supplies. Companies that use the latter reprocess the water using methods such as distillation, reverse osmosis, de-ionization, and filtration.

US law requires all drinking water to be treated so it does not go bad in the bottle, Boyes explains. Since there is not a taste difference, all non-carbonated water is basically the same.

And that's where branding comes in. (especially for high end products)

"Branding is extremely important for water," says Chiranjeev Kohil, professor of marketing at California State University at Fullerton. "In a lot of categories, you can duplicate products and get an edge on quality or attributes, but that edge can be shaved off very quickly by competitors. In the water category, there is no technological superiority. The only thing that differentiates one water from the next is the brand."

Marketing experts say that a commodity such as water can be branded effectively. The key is differentiation, but that is also the challenge. After all, water is water.

"The source of the water provides a good foundation for differentiation," says Rob Frankel, a branding consultant. "Evian is from the Alps, Arrowhead is from the mountains in California, etc. If you are going to tap into the brand culture, so to speak, you start with the singularity of the source."

Harvey Briggs of Lindsay Stone and Briggs recommends that marketers "grab the high ground" and focus on key benefits that consumers are looking for. "The people who grab, say, purity and own that with the brand are the ones who are going to succeed in the long term," says Briggs, executive vice president of the brand innovation firm in Madison, Wisconsin.

If there were not any branding in the water category, the business would eventually go 100 percent to price and to private label, says Robert Lynn, executive vice president of sales and marketing for Global Beverage Systems, marketers of Le*Nature's Beverages. For the time being, he believes that brand is more important than price.

"Price won't become the major factor until growth stops," he predicts. "When the growth stops, that's when the war starts, and you fight the war on price. How do brands as strong as Coke and Pepsi fight their wars? On price. Two-liter soda for 69 cents. And those brands are stronger than any of the water brands."

US Brands
The US bottled water market has been reshaped in the last few years by the entry of Coke and Pepsi and their respective brands, Dasani and Aquafina, both filtered tap water.

"Each of these two marketers has introduced a single product that has been heavily promoted and enjoyed the support of the best distribution systems to be had in the beverage industry," according to a report on the US bottled water market by MarketResearch.com.

Pepsi's Aquafina, introduced in 1997, is now the number one branded non-carbonated bottled water in the US. Coke's Dasani, launched a few months later, is second in the category. Both are likely to lead the market in the future.

"When you become a brand leader, it's not only because you had a good, strong name," says Kohil of California State. "It's also because you had a very efficient distribution system in place. That is the advantage that [Coke and Pepsi] have. They already have the relationships with the distributors and retailers."

Meanwhile, the regional brands must rely on local loyalty as they compete "against the big, faceless multinationals," says Briggs of Lindsey Stone and Briggs. For example, consumers in New England may be inclined to buy local natural spring waters such as Hidden Spring or Twin Mountain from Vermont.

"The regionals will succeed through grass roots efforts," Briggs says. "It's the things they are going to do to get closer to their communities that are going to differentiate them."

Frankel, the consultant, believes that store brands will account for most of the business because they ride the advertising coattails of the major marketers. Consumers will support the category but spend less because they are buying private label.

The category will see its share of new products in the form of enhanced waters that provide energy, promote fitness, or just taste better than plain drinking water. Recent entrants include Propel, a purified water beverage with vitamins from Quaker Oats, the makers of Gatorade; Fluoride To Go, a fluorinated spring water from Dannon; and Fruit 2 Go, a naturally-flavored spring water beverage from Very Fine Products.

But the war will be in non-carbonated bottled water. Market analysts look for major consolidation among the plethora of brands in the next few years. Essentially, the large national marketers will buy local brands around the country and shut them down. Why? To reduce competition and, in some cases, to acquire other supply sources for spring water.

High Profile War

The battle between Coke and Pepsi and the larger European brands is the "high profile war that is being waged," predicted Briggs, who adds that **branding will remain a deciding factor** for discerning consumers. "Quality and trust are going to be critical, so brands will be important."

Ric Davidge, MPA

6. Why Alaska *natural* Glacial Water now?

6.1   Why water from Alaska, and why twenty sources to start?

Source Security
First it is critical to understand our rare international 'source security'. That the water rights/property rights we receive in Alaska are globally, incredibly unique. Even in a future water emergency, the state could only 'take' our water even by court action with market-price based compensation and then most of our sources are nowhere close to a community that would need it. The state would basically have to pay us for our water and/or their use of our infrastructure. We have been careful in our source selections. Given the value of our water, once developed and in market, government acquisition is not likely just based on the cost and their alternatives. In addition, our Alaskan sources are in one of the most productive freshwater regions in the world and a shortage here is not likely. The longest draught recorded in southeast Alaska was seven days. With annual precipitation of 200 to over 300 inches a year, this is a very wet area of the world that few humans know or understand.

None of our selections in SE Alaska have any water connection to or impact from Canadian mining. This is important because there have been periods in the past where Canadian mining left contamination in waters that ran in to Alaska. We bring this forward because we have been asked this question by some who have expressed interest in our bulk export potential.

As we have pointed out the brand, ***pristine natural Alaskan glacial water***, and its unique top-of-mind marketability are the foundational marketing reasons it makes sense to start with sources in southeastern Alaska. The other part is private ownership and their proximity to markets.

Jones Act Restrictions
It is important to remember that we still cannot ship water into California because of an old federal law to protect marine unions, The Jones Act, which prohibits any foreign vessel from conveying anything – including people and water – between two American ports. This may change soon but based on our

work both in California and in Washington, D.C. it is not likely likely until after the next national election – if we win. The cost of a vessel that must be owned, designed, constructed, and operated by Americans - to haul sufficient water to make it work financially is at least 30% greater than the same construction in South Korea and the O&M cost with foreign vs American crews. If an American company can compete in ship building, we are interested because that opens more very thirsty markets in close conveyance from Alaska making these markets very attractive. But none yet.

AQUEOUS International, Inc. is and has been involved in carefully assessing the high-volume V Maxx tanker accessible to *natural* glacial water sources in southeast Alaska for years. These are raw natural freshwater sources <u>with no salmon</u> that can provide up to 90 million gallons a week from each source for export. With two or three buried vinyl pipes our design will load our freshwater V Maxx VLCC class tankers in 30+ hours. The water is extraordinary and drinkable **without any treatment, although we will oxidize in transit, it will remain** in its *natural* state upon delivery and its association with 'Alaska' and 'glaciers' greatly enhances its bottled water or base stock for other products marketability.

Just think of a cosmetic product that uses our water on the shelf of a major retail market anywhere. Our brand names scream out that 'we are different' better – try us.

Secure Contract Law

Water export contracts involving sources and municipal/state governments in the United States are **the most secure**, under the rule of contract law, in the world. It is our preference that delivery contracts be for a term of 30 years minimum (the projected life of a new V Maxx) and we are extremely interested in 99-year contracts – some exist in the Mediterranean but use barges. But remember most are controlled by organized crime families. This is the 'generational' nature of this business and one of the very reasons we have been persistent in our work in all areas of water in the world.

Every market partner, as part of our sales agreement, is flown into SE Alaska and shown, by plane, the water sources specific to their delivery contract. Marketing videos and still photos are also provided to the partner/buyer.

6.2   Our Southeastern Alaskan Sources

Twenty Potential Southeast Sources

We have identified and assessed over twenty potential **low development cost** bulk water sources in southeast Alaska. We do not reveal the names or locations of these sources until we have secured what is called 'under paper' protection of each site that does not allow anyone to over claim our sources. We will receive and retain actual 'ownership' of the water appropriated for each site at the close of the process. This is important in protecting our investments to date and any future investments. All twenty of these sources will be legally 'under paper' and most sources will be fully appropriated 'owned' before the end of the first twelve months of our operation in southeast Alaska we anticipate in 2021. Again, we only actually develop a source with infrastructure when we have a **signed delivery contract with a down payment.**

Work to be Done

Our experts in hydrology, glaciology, fisheries biology, wildlife, forestry, archeology, oceanography, engineering, surveying, tanker pilot/access and egress, etc. are a team with unprecedented Alaskan experience and competence in our effort to secure every one of these sources in less than 12 months. This includes site inspections with many of these experts flying in weather that is not often inviting. But we have and will continue to get it done. We have secured the best sea/freshwater engineering company in Alaska that Mr. Davidge has worked with in many areas of the state on a wide variety of projects. We have secured the support of key politicos to ensure we have no problems in that arena. We have already interacted with the reginal commercial fishing fleet in southeast Alaska to help them understand who we are and how we will interact with them so as not to harm their business. We know the communities around which our sources are located – some at distance. We have lived in some and worked in others. The bottom line is we have done our homework – but still like to learn. And we continue incredibly positive negotiations with the largest Alaska Regional Native Corp in SE Alaska in a move that solves many possible issues in the future.

Every source must have the capacity, with our and the state's protocols, to provide 90 million gallons a week every week, every year (summer and winter) for infinity, but allowing two weeks for stuff that may happen. This is a generational new industry for not just Alaska, but other areas of the world we have explored. We are ready to get back into the wilderness and finish this important step that protects us and our investors for lifetimes.

But why 20 sources?

We must be able to adjust to weather, 30-foot tides, minus tides, commercial fishing, etc. as our tankers approach. Each tanker calls in to confirm the load location and/or is directed to an alternate in the area if necessary. The number and their groupings allow us these options so that we are <u>always on schedule</u> for every market partner. It just means we will always be on time at market.

- Adequate natural discharge to allow annual harvest of up to **4.5 billion gallons <u>per source</u> per year**. With twenty sources that is **90 billion gallons a year - max**
- Accessible, with turn radius, by our custom designed freshwater V Maxx tankers
- Provably do not have salmon in their discharge (a key requirement by us)
- Water source surface area large enough to ensure clean recharge with no likely possibility of any development that could harm water quality. These are **very** remote areas.
- At tidewater, with a water discharge at <u>sufficient grade</u>, do not allow any anadromous fish access.
- Environmentally viable and sustainable year-round, as seasons very in system discharge
- Culturally acceptable to First Alaskans in the areas who have feelings about 'their' water
- Technically viable with manageable operational and maintenance costs
- Adequate space and supply services for each site's caretaker who may be rotated every two weeks, as agreed in each contract (some may not want to leave as often)
- Ensure no one other than AQUEOUS, International, Inc. can claim access to or use to these waters.

We are already in the process of filing the request for bulk water export appropriations for some of these sites consistent with the laws of the State of Alaska. Each application costs:

- $1,000 application fee for water rights (ownership)
- $50 per hour of State staff time to adjudicate.
    - Estimate $2,500 to $5,000 in state fees for full adjudication of each source.
- Other Costs
    - Hydrologic and Environmental Assessments (underway)
    - Tanker access formal confirmation (tanker pilot)
    - Agency Reviews/comments/responses (30 days)
    - Public Process/responses (30 days)
- Once we are confident, we can secure the state's appropriation to our ownership we do pre-development cost estimates with (PND Engineering). This is already underway for some sources. We do not plan to build anything until we have a signed Delivery Contract, with a deposit, that justifies the investment. We have sufficient time as a VMaxx tanker takes 23 months to build if we are responsible for the conveyance to market.
    - Develop a pipeline route (2 or 3 1/2-inch vinyl seamless pipes) from source to a small pump house at ordinary highwater mark/tidewater. All sources already have access to electricity.
    - A caretaker home/workstation. (Caretakers are independent contractors) Includes regular (3 per week minimum) drone flights over the source and its recharge zone and any areas of concern with video kept on computer and sent to the main office as necessary. If there is any concern, the main office is immediately notified and the caretaker visits the area of concern, photographs it, fixes it, and reports back. Most likely areas of concern deal with wildlife interacting with the water source.
        - The caretaker rotates every two weeks (or as agreed in contract)
            - Fuel storage if needed (we prefer electric heat)
            - Fuel storage for vessel on barge

- Crew vessel (small work boat safely able in local seawaters)
- Support services (food, fuel, etc.)
  o Design and set pipe from pump station to buoy (subsurface)

Five Year Budget

For more details on site development, review the working five-year budget presented later in this plan. Site plans must all be custom designed given the location of 'take point' to ordinary high water, grade which can be significant, and general uniqueness of each site. This information is only shared with investors, market partners, and state permitting personnel. We have already had very good meetings with the Commissioner of DNR and the key people who will be involved in permitting our sites. We do not give anything to any media – ever.

First in Time, First in Right

We have chosen 20 sites that meet a range of business, environmental and regulatory criteria. Each will provide up to 90 million gallons a week or 4.6 billion gallons a year for export aligned with state law requirements. The actual cost to fully develop each site for export will depend on the unique specifics of that site and estimates are shown in the enclosed proforma after engineering preliminary consulting. A final design is required by the state and will be submitted to the state prior to its full appropriation of the water right requested and the issuance of the export license and other permits, but our *'first in time, first in right'* status already legally protects us from any other application for water or other use from our sources. The Department of Natural Resources will review the preliminary design and provide comments on that design so that our final design is accepted without delay. Our engineering selection is highly respected in Alaska and by both state and federal agencies.

Site Acquisitions Under Way

The acquisition of our twenty initial sources is already underway. We have some sources 'under paper' now and others in process. All twenty sources will be owned by the company within the first 12 months of receipt of investor funding. Thus, the company has a very marketable asset (privately ***owned*** high end natural Alaskan glacial water) for leverage. This level of 'ownership' is rare on this planet. To help understand the complexities and choice options during the site/source acquisition process we have developed a detailed process diagram that is provided on the next page.

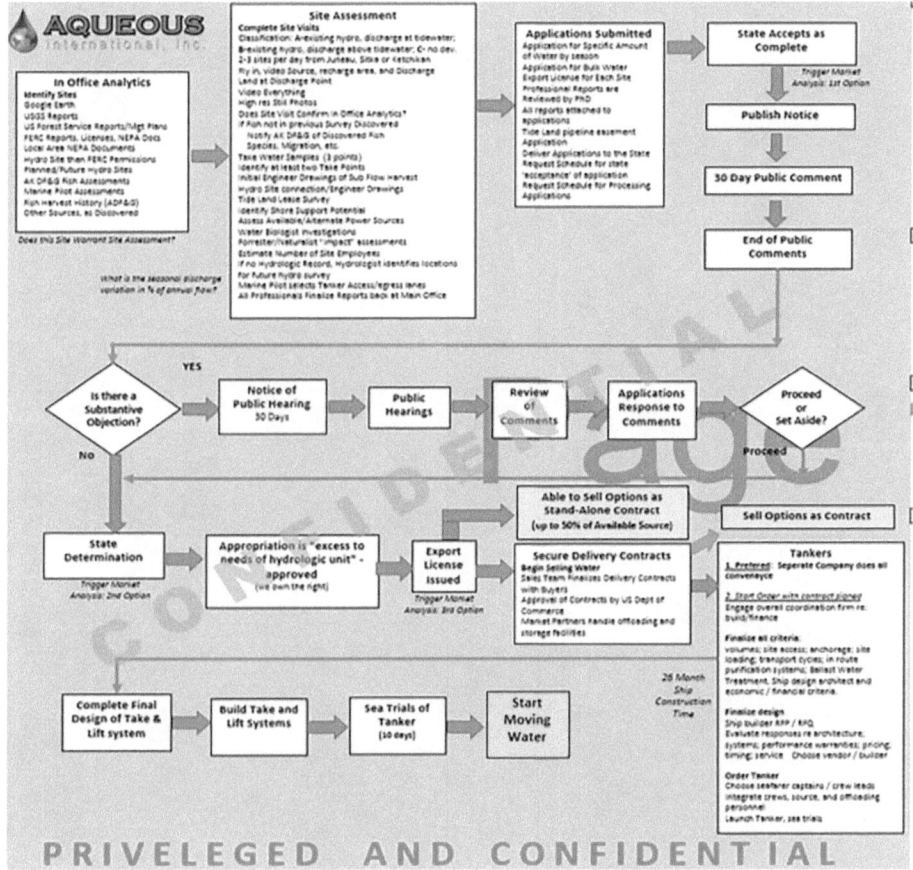

Figure 29: Development Process Flow Chart

7. PRODUCT DESCRIPTION

High Quality Natural Water
This venture is initially structured to eventually provide up to **100 billion gallons** a year of exceedingly high quality *natural* Alaskan glacial freshwater in bulk quantities with the eventual potential of delivery of 90 million gallons a week per site per market and more. Other sources have been identified and can be developed as delivery contracts warrant. All sources are secured with full corporate private property ownership by AQUEOUS International, Inc. These assets provide significant financial leverage and attraction to potential commodity futures, now in development.

The water sheds of all sources are uninhabited and protected by federal, state, and local laws from any development (ranching, farming, or any other activity) that could potentially compromise the quality of the water. The water quality is continuously monitored by the company with likely inspections by the Alaska Department of Environmental Protection and the Alaska Department of Natural Resources – likely once a year. These sources provide some of the purest oldest (some 10,000+ years) natural glacial freshwater on earth.

Continuous Water Testing
Each source will be tested **with each lift** as it loads a tanker. This test is immediately sent to our main office and is then provided to the market partner enabling the highest confidence in the quality of the water they have purchased and will soon receive. This water will be maintained in transit to markets using UV circulation treatment and oxygenation to ensure its freshness upon arrival. **It will NOT need to be filtered or otherwise treated** at market. The water will also be tested upon arrival at market and that test provided to the AQ main office and then to the buyer.

Water in the United States is measured and sold wholesale by the thousand gallons, or acre feet. In the rest of the world, it is generally measured and sold in cubic meters. Conversions have been made in this plan to assist the reader in understanding the amount of water being discussed. It is generally easier if we discuss water in the 'gallon' rather than cubic meter measure as that is much closer to the wholesale and/or retail scale used in almost all markets for this type of high-end 'fine' water.

Expansion

The site visits provided to every market partner to these, up to 20 sites, allow them by **direct personal experience** to speak to the value of these pure natural Alaskan glacial waters and the capability of what we have presented in this document. We remain open to expansion, especially if California is successful in exempting water from the Jones Act which will allow us to ship to America. Most really have really no idea of the vast extent of our waters. To date no one has accurately characterized the total amount of water discharged into the Pacific from this hydrologic region. Yes, there are estimates, but this is a venture we may someday take on.

We recently flew a member of the Colorado River Commission over these waters. He was born and raised in the deserts of Arizona, and he actually and embarrassingly broke down and cried as we flew over water sites. He stated that he would never have believed it if he had not seen it firsthand. More here when it develops.

Water Conversion Table

Bulk Water is sold by the acre foot or 100,000 gallons in the United States. In the rest of the world, bulk water is sold by the cubic meter. Because the initial market for this water is bottled water, we will refer to it in gallons. We will also use a $.01 as a place holder for cost/price/profit until we have actual delivery contracts that set these cost/price numbers per market partner.

Acre Feet/Cubic Meters – etc.
- 1 AF = 1,234 cubic meters = 325,851 gallons = 43,560 cubic feet = 1,089 tons
- 10,000 AFY = 12.34 million cubic meters = 3.3 billion gallons (3,258,510,000)
- 20,000 AFY = 24.68 million cubic meters = 6.6 billion gallons

Cubic Meters 1 cubic meter = 1 ton = 264.17 gallons
- 10,000 cubic meters = 8.1+ acre feet
- 20,000 cubic meters = 16.2+ acre feet
- 30,000 cubic meters = 24.3+ acre feet
- 50,000 cubic meters = 40.5+ acre feet
- 100,000 cubic meters = 90+ acre feet

Gallons (US liquid)
- **100 M gallons = 378,541.178 4 cubic meters**
- **84 M gallons = 317,974.589 86 cubic meters**
- 50 M gallons = 189,270.589 2 cubic meters
- 20 M gallons = 75,708.235 68 cubic meters
- 10 M gallons = 37,854.117 84 cubic meters

Tons
- 1 ton = 1 cubic meter = 264.17 gallons
- 10,000 tons = 10,000 cubic meters = 8.1+ AF = gallons

Meters
- 1 meter = 3.28+ feet = 39.37 inches

Water Transport Bag Sizes (largest tested)
- 50,000 cubic meters/tons   length   260 meters or 853 ft
                              width   30 meters or 98.5 ft
                              depth   7 meters or 24.5 ft

Beverage Sizes
- 1 gallon = 3.8 liters (3.785 411 784 liter)
- 5 gallons = 19 liters (18.927 058 92 liter)
- 2.5 gallons = 9+ liters (9.463 529 46 liter)

Figure 30:   Water Conversion Table.

7.1   Water Quality

With no development, ranching, farming, or other activity conducted on their freshwater shores, our sources offer pure water in its most *natural* state. The water, therefore, does not require any **filtration** or disinfection prior to distribution to consumers. However, to fully ensure that we do not have any concern, our freshwater tankers will treat the water using ultraviolet light and an oxygen freshener during conveyance. All water taken will be tested at lift and it will be again tested upon arrival at market prior to the offload so that our market partners and others have the highest confidence in the quality and uniqueness of our water.

Thus, bulk water for export from our sources will, the federal EPA and state DEC drinking water standards without filtration. This is an extraordinary marketing and production advantage few other *natural* sources on the planet can claim. The other marketing advantage is the relationship of this water source with Alaskan glaciers in ages of between 10,000 years plus. There are glaciers as old as 32,000 years, but not in our plans.

Each year AQEUOUS International, Inc. will submit a **Watershed Control Report** to the Alaska Department of Environmental Conservation and the federal Environmental Protection Agency – both of whom hold water quality for human consumption powers over the protection of such water sources. Although there remain questions regarding the legal authority of the federally required reports because the sale of the water is to a private owner and then a foreign market, AQUOEUS International, Inc. has chosen to provide these annual reports as part of their commitment to ensure pristine *natural* glacial freshwater to all markets regardless of their national origins. These reports address the following:

1. Any special concerns regarding the watershed and our handling practices
2. Outlining any activity that could adversely affect the water quality in the watershed
3. Any planning for future activities that may have an adverse effect

on the watershed (certainly not something we would support)
4. All water quality reports (at load and at off load) and summations on specific markets the water is sold to
5. Present the certification statement and signatures

These reports, along with the federal and state approvals, are then sent to our market partners and others as appropriate annually so that they are available for marketing and to ensure buyer confidence in our product. Marketing then can report having these federal and state approvals for our unique brand.

It is the quality, the naturalness, the association with glaciers, and Alaska of these waters that makes them so marketable as a high-end bottled water or product base in the markets targeted.

7.2 What we provide to markets

"Ready to Drink" Water
Alaska's natural glacial water is an exceptional raw natural freshwater for bottled water or beverage markets. The fact that it does not require any filtration or other treatment prior to use (it will still receive a UV treatment in transit for insurance) is not only market attractive but also margin attractive.

One of the main tanker design considerations for us is the need to deliver this water "ready to drink". It is economically viable that while the tanker is in route to China or India that we run the water through a UV system. This greatly enhances the immediate market value of this water without adding any chemicals to the water at point-of-sale by reducing the cost to the buyer for any treatment prior to bottling. We believe the V Max VLCC specifically designed to lift and treat water with UV is very cost effective and significantly enhances the "value" i.e. the asking price for the water at market.

Natural Minerals
The other value of this natural water is that the addition of minerals will

not be needed for taste. This is a growing additive to bottled water as consumers learn of the health value of natural minerals. Some waters, after filtration, treatment, or distillation have a very flat taste and many bottling companies must add supplements such as minerals to enhance taste and thus consumer demand. This of course adds cost to the product and lowers margins.

The marketing value of pristine *natural* Alaskan glacial water in such markets as China and India will almost certainly change the "shelf habits" of buyers in these markets. The two terms, "Alaska" and "glacial" are significant top-of-mind buyer attractors. Shelf tests in the United States conducted some years past showed a high propensity for buyers to reach for the Alaskan water products as their first preference if costs were the same or close to other brands.

Alaska Perceptions
The advantage of a well-established, almost intuitive "top of mind" perception of Alaska and its Pleistocene (10,000 to 30,000-year-old) glaciers as pristine is an extraordinary marketing advantage over any other bottled/packaged water product on the shelf – even Evian or Icelandic.

This also works for other beverages or food or even manufactured products as feed or base stock. If you sell a flavored beverage built on *natural* Alaskan glacial water or Alaskan 'ice' water, that beverage will achieve a marketing advantage topping its competitors.

Key Marketing Item
One of the keys in the marketing of this unique water will be the bottle design and labels used by those who receive this water for their buyers. It may be in the best interest of this venture to partner with an international marketing firm who understand this opportunity and can work with our buyers in creating unique bottles and labels for this product. We are extremely interested in assisting any market partner in this area as we have decades of experience and experienced professionals who do this for a living.

High-End Product Marketing

One of the keys in marketing mid to high-end or "fine" bottled waters is to market the water's source. We do not want to show tankers, but we do want to show the pristine *natural* area of the water's source. Therefore, during our summer site assessment work, we take our Marketing VP and business development consultant and our professional video/still photo team to every source so that we have high quality still and video images that they can use for digital development of marketing images. It is interesting that even with these images, investors and market partners 'need' to see them personally – to believe and grasp a sense of the magnitude of what we are doing as well as the security of our sources.

Public Awareness of Contamination

The other areas we can aid markets are in developing public/market awareness of freshwater contamination in key areas and how to leverage that 'consumer fear' in getting the consumer to choose pure natural Alaskan glacial water - their preferred choice. Because we monitor all water contamination developments globally, we can bring immediate attention to new information that can help quickly leverage consumers to buy our product. It also allows us to help in the development of unique products for subsets of consumers. These can include water for pregnant women, men who are concerned with the strength of their sperm, newborn children, the aging of both men and women, and more.

8. BULK CONVEYANCE OPTIONS

Conveyance Separately Owned and Operated

AQUEOUS International, Inc. is open to various bulk conveyance business structural options. We think it makes good business sense for the conveyance systems to be separately owned and operated, with some participation by AQUEOUS in the design and testing phase, by a separate company. After all we are the global water experts, not the conveyance experts.

The information provided here is what we have learned as we have struggled with the conveyance questions, access, and cost, for bulk freshwater over the past two plus decades. What we think works for our markets and what does not – but may need to be considered for other markets in the future.

There are basically four conveyance technologies that may be able to move bulk water in adequate quantities from southeast Alaska to transoceanic markets. The choice and application of these options is generally based on access to sources, cost at sources, distance to market, and lift capacity that enables essential water harvesting and conveyance economics.

8.1   Barge

The use of a **barge** configured to hold potable water will also require a sea tug. This option has inadequate scale/lift, energy cost, and slow speeds are the limitations due to distance and potential weather across the Pacific. At present this is NOT an appropriate conveyance technology for **this** venture. Tugs and barges are the current mode of operation in most of the Mediterranean even today, but their markets are relatively small such as NATO facilities on islands and small coastal towns. Desal in this area is prominent but not popular due to the growing knowledge of its adverse health effects and the growing cost of maintenance and required upgrades.

8.2 General Bulk-Carrier

A **general bulk-carrier** fitted with a water bag or liner within the ship's hold. This is an option for a used (such as one formerly used to convey oil) tanker depending on scale/lift capacity, but if a tanker is going to be converted to carry water, far more than just the hold needs to be lined or replaced. Essentially **all surfaces that have ever touched oil** that will now likely touch water will need to be replaced, lined, or "cleaned". The cost of almost any cleaning requirement is extremely high and studies by such firms as NYK Japan have found that full conversion of a used tanker is not efficient over the remaining life of that tanker. Crude oil tankers have a sea life of 15 years, and even if converted to freshwater after previously lifting crude you may only extend that life by 5 years. This is not a recommended strategy for this venture.

8.3 Tanker Ship

Preferred Alternative

A **tanker ship (standard design)** specifically built or converted but has had no contact with petroleum products to lift potable water is **the preferred alternative**. There is concern with its draft, given the access (depth and turnaround) limitations to some of our sources. Depending on the distance (time) to markets and water temperatures, some type of water quality management or enhancement, we believe, is needed on these vessels to maintain the freshness and integrity of the water. A standard designed tanker to lift freshwater, that has never lifted crude, can realize a sea life of maybe 20 years.

Large tankers are designed to bend in the sea causing stress cracks to appear over time in the steel hull – that must be repaired. The purchase or lease of a new uncontaminated tanker, given the required lift capacity/scale and cost, we recommend we start with a VLCC class tanker and continue using that class of tanker over the life of this venture. If we develop alternative water sources in Russia and/or Albania for example (we have some) we may want to consider other tanker sizes such as Panamax or Suezmax, but for this plan, the

VLCC class is the appropriate scale of tanker economically. This application is only economically viable if the end user can place the water received into a bottled beverage without additional treatment or filtration.

8.4   PREFERED Tanker Option

The **V Maxx VLCC** class tanker design by Stena Bulk of Sweden is the most appropriate conveyance for this venture. The V Maxx modifications increased lift from 84 million gallons to 90 million gallons for freshwater (oil can fill at 92 million gallons because it is lighter than freshwater), increased speed by 2 knots (which we now know can be further enhanced with forward projected bubble technology), took 16+ feet off the draft requirement (big deal) allowing greater and much cheaper access to our sources, and allow a turning radius with forward and aft props that is dramatically smaller than the standard VLCC tanker. Although the V Maxx was designed by a Swiss company, they are most often built in South Korea due to labor and material costs. The quality reports on the S. Korean (Hyundai Heavy Industries) ships are excellent. We will meet with them soon to outline our design requirements for pristine water. More detail on the V Maxx follows.

8.5   Water Transport Bags (WTB)

The use of large poly-fiber bags known as **Water Transport Bags (WTB)**. These bags, developed for us to use in the Mediterranean, are towed behind a modified sea tug. (see video on our website) This technology has been proven by AQUEOUS International, Inc. in the Mediterranean while a partner in World Water SA in between markets from Turkey and Cyprus, and several Greek islands. They have NOT been proven in the open Pacific or Atlantic seas, however new aft bag steering technologies may assist to increase their speed and tail stability in such transoceanic applications.

WTB have been designed and used in a variety of sizes to include 35,000 to 50,000 cubic meters. Given a specific market, the size of

WTB is determined by the market's seasonal demand and adjusted by its frequency of delivery. Obviously the larger the bag the better the cost efficiency, but you must also take into consideration the nature of the seas to be traveled. Bags have separated from their tugs and traveled across large areas before being recovered causing significant and expensive navigational hazards to other vessels with unanticipated costs to their owners. Additional safety systems have now been designed and real-time computer and GPS monitoring of larger bags developed to assist in the monitoring and recovery of separated WTBs.

Sea Storage

WTBs are a proven conveyance technology – but they can be more. They can also be used as "in sea" storage by anchoring them in the sea at market. There may be some application in southeastern Alaska for bulk water sources that do not accommodate V Maxx VLCC tankers due to depth or turning radius. Bags could be filled and then towed to a central location to be pumped off into tankers for cross sea markets. We carefully avoid this added cost, but such an application could easily open several new water sources that otherwise might not be available. The problem is additional operational cost. The rule of thumb is that every time you change conveyance you double the cost, but this can be explored if it makes sense given the scale of a source and distance to a tanker and/or buoy.

Example: WTB @ 20K cubic meters which was used for fabric and other testing.

Figure 31: WTB Testing

8.5.1　Loading

The use of tankers (Panamax Class, Aframax Class, Suezmax, or a Very Large Crude Carrier) is most often made available by an underwater piping system from the terminal head to an area of the bay with sufficient water depth and surface stability to accommodate the ships berth and turn around area. These carriers can lift between 20 to 90 million gallons of water per trip.

8.5.2　Panamax Class Tankers

A typical Panamax Class tanker depending on age (60,000 to 70,000 DWT) hauls about half a million barrels, or 64.5-acre feet, or **21,000,000 gallons** per lift.

Figure 32:　Panamax Class Tankers.

Panamax ships are the largest ships that can pass through the Panama Canal. The size is limited by the dimensions of the lock chambers and the depth of the water in the canal. An increasing number of ships are built to the Panamax limit to carry the maximum amount of cargo through the canal.

The increasing prevalence of vessels of this maximum size is a problem for the canal as a Panamax ship is a tight fit that requires precise control of the vessel in the locks, often resulting in longer lock time (2 to 3 days), and requiring these ships transit only in daylight. Because the largest ships travelling in opposite directions cannot pass safely within the Gaillard Cut, the canal effectively operates an alternating one-way system for these ships. This then translates into as much as a 3 day wait on either side. So, you are looking at possibly 5+ days to get through the Panama Canal along with fees. This may make this choice not economical for many markets.

The maximum dimensions allowed for a ship transiting the Panama Canal are:

- Length: 965 ft (294.13 m)
- Beam (width): 106 ft (32.31 m)
- Draft: 39.5 ft (12.04 m) in tropical fresh water (the salinity and temperature of water affect its density, and hence how deep a ship will float in the water)
- Air draft: 190 ft (57.91 m) measured from the waterline to the vessel's highest point
- Speed is on average 13.5 knots
- Lock time is up to three days
- Wait time either side is up to three days
- Day cost, including fuel is about $40,000 or more a day

The Panama Canal Authority also charges $250,000 <u>or more</u> in fees for Panamax tankers. These fees are not controllable by this company and can be change at any time by the Canal Authority, **which is a political entity**. So, this becomes another volatile operating cost in our effort to consider exceedingly small margins.

There is an ongoing effort by the canal authority to improve width/length canal limitations and transit time for larger tankers. When this will occur is anyone's guess, but it is likely to happen based on the pressure of the global shipping industry which is now changing due to the opening of the arctic to international shipping.

A Panamax cargo ship would typically have a DWT of 60,000-70,000 tons and a maximum cargo intake of 52,500 tons. Depending on age, capacity in barrels is 419,000 (13,198,000 gallons) to 527,285 (18,499,477.5 gallons). As you can see these numbers change depending on the specific vessel. If you can also use ballast tanks for freshwater lift the capacity can improve by up to 20%.

8.5.3 Aframax Tankers

An Aframax ship is an oil tanker smaller than 120,000 metric tons deadweight (DWT) and with a breadth above 32.31 m. Aframax class tankers are largely used in the basins of the Black Sea, the North Sea, the Caribbean Sea, the China Sea, and the Mediterranean. Non-OPEC exporting countries may require the use of Aframax tankers because the harbors and canals through which these countries export their oil are too small to accommodate very large crude carriers (VLCC). The term Aframax is based on the *Average Freight Rate Assessment* (AFRA) tanker rate system.

Figure 33: Aframax Tanker.

The Aframax tanker would allow shipment to the Middle East without going through the Panama Canal and present 22 million plus gallons to this market. There is a large and growing inventory of this class of tanker and we may be able to acquire one <u>before</u> it is contaminated by oil or chemicals.

The average dimensions of an Aframax ship are:

- Length: 243.2 m
- Breadth: 41.6 m
- Draft: 14.1 m
- Barrel Intake: 690,000 bbl.
- US Gallons: 21,735,000.001 gallons
- Speed: 13.5 to 14.7 knots
- Day-cost including fuel is like a Panamax at about $40,000 - $50,000.

- Bear in mind that these are *average* figures.

Having tankers that can carry almost 22 million gallons of freshwater at a speed of 14.5 knots, without paying the Panama Canal fees or sitting idle for up to three days appears to offer some efficiencies to this project. We are exploring these options for some markets.

8.5.4   VLCC Class Tankers (not a VMaxx)

A typical VLCC Class tanker hauls 200,000 to 325,000 DWT with capacity between 60,500,000 gallons and 84,000,000 gallons. Again, the age and manufacturer can make a significant difference in the lift capacity of these tankers and as stated before, using ballast tanks, if feasible, can increase the marketable water lifted by 20%. But remember these ratios are based on oil, not much heavier freshwater. This is the class of tanker most likely to meet our immediate needs for markets like China and India if we need to move 'now' for a specific market that cannot wait for the 23-month construction of a V Maxx. A cost table is enclosed later in this plan that outlines the cost and profit potential of a new VLCC tanker already under consideration.

We continue to look for an appropriate freshwater tanker designer and ship builder for this new type (freshwater) of tanker. We have made contact with

Hyundai Heavy in South Korea who have built VLCC class tankers. This is how we discovered that there are VLCC class tankers in development but not contaminated with oil – yet.

8.5.5 VMaxx VLCC class tanker (preferred)

In the beginning of 2010, we became aware of a new VLCC class design by Stena Bulk, in Switzerland known as the VMaxx. This discovery was extremely exciting for a range of reasons.

The double hull of the VMaxx has a low molded depth of 25.6 meters – around 5 meters or over **16 feet less** than that of a new conventional VLCC. This design change opens many additional sources in southeast Alaska for lift. Oil cargo intake at 16.76 meters (55 feet) is 75% higher than a standard Suezmax and 20% higher than a standard VLCC. Remember freshwater is heavier than oil.

The VMaxx – built for at active in at least 25 years of trading when hauling crude oil, may have an **extended life of 30 years when only lifting freshwater** – is thoroughly protected against corrosion for extended fatigue life. The ballast, cargo, fuel oil and contaminated water have ballast tanks or void space between their boundaries and the hull. As an extra precaution, in the event of grounding, the hopper tanks are larger than on a conventional VLCC to reduce the risk of an oil spill. With freshwater this is not a concern, other than ship fuel. But, since we do not carry crude, the larger tanks offer higher lift capacity.

8.5.6 The V Maxx:

1. Can increase our lift capacity by maybe 30%
2. Lowers our fuel consumption per gallon
3. Extends the life of a tanker to **30+ years** when hauling freshwater. That is 10 more years than a standard VLCC tanker. This is huge in ROI.
4. Can travel 2 knots faster than conventional VLCC with no greater fuel consumption (due mainly to drag reduction). Other new drag reduction technologies, such as bubble discharge forward may add another knot or two and decrease energy demand.
5. Because of its shallow draft (16+ feet less), it allows this tanker to access far more water sources not only in Southeast Alaska but in

many other areas of the world.

6. Has the latest design and capabilities in navigation, control, turning radius is significantly reduced - an especially important issue for some sources, etc.

The **V Maxx VLCC class tanker is the most attractive conveyance system for this venture** because their size (depth & turning radius) allows them to access more sources in southeast Alaska and provides sufficient lift volume and cost containment to make them economical for our purposes. Standard VLCC class tankers can carry up to 84 million gallons of water per trip, with a turn-around or round-trip time of 36 days to China for example, depending on vessel speed, water source location, delivery point, and load or off-load capabilities. The VMaxx design allows a lift of 90 million gallons of water and adds two more knots of speed along with a turning radius that may be necessary to access/egress some sources in southeastern Alaska. These tankers are uniformly outfitted with large pumps, allowing them to be pumped off within 30 hours depending on the receiving size and number of the pipes at tidewater at market. With the VMaxx, roundtrips from southeast Alaska to China is about 31 days. This is a critical consideration when negotiating a Delivery Contract as time is money when a tanker is sitting at dock. Turn around has also been adjusted with the upgrade of the VMaxx allowing it to turn on a quarter using forward and aft turning systems. This again opens new sources to us.

Suez Canal

Another new discovery is that a VMaxx can pass through the Suez Canal. It is tight, but it works. With the water sources we know in the Atlantic artic and Mediterranean region, this could be a significant advancement for the company in this very thirsty market.

At these volumes of lift it will take safely 3 days with margin to fill VMaxx VLCC tankers based on the current design capacity of the pipes/hoses at tidewater. That is a total of under 5 days per visit (3 to fill, one to come in and one to depart w margin).

Building Priority

We continue to explore this development with the principal companies who designed and are currently building and operating this design to ascertain day-costs, etc. We have made formal contact with the CEO of Sterna Bulk and were in discussions with their designers, when Stena got an offer, they could not refuse and sold all their V Maxx to a company in Europe. If possible, for our China markets, we would like to have this design built in South Korea, where Hyundai Heavy has been building V Maxx for years, and their labor costs are much less then Sweden and their quality outstanding. This is a top priority once we have a delivery contract.

There is always something in the tube that may come into the conveyance market that we do not know about even though we track these developments as best we can. We do know about many options that are not worthy of consideration. We have often found them scams to leverage initial investments. But we are in touch with this industry in an ongoing strategy to consider any responsible option. For example: we were recently advised of several 'juice tankers' going out of business. Most in the Atlantic, but after some serious work we found the cost variances and the 'proven' lift and travel costs to not make them viable for our interests. We continue to listen and inquire.

Figure 34: ULCC Class Tanker.

This is one of the largest and newest ULCC class tankers. *Hellespont Alhambra* (now *TI Asia*), a ULCC TI class supertanker. We present this photo to show how clean and dramatic a white tanker appears against the sea. **This would be our preferred vessel color with aqua blue lettering and trim.**

8.6 ULCC Class Tankers

(provided only as a point of comparative information)

A ULCC Class tanker can haul up to 300-acrefeet or 407,049 cubic meters, or **107,530,972.9 gallons**

At this point we do not believe ULCC class tankers can access most of our sources or any other potential water source in southeastern Alaska. Additional research will be required to find water sources for this class of tanker. The problem in most ports for ULCC tankers is turnaround space, not draft. The ULCC requires about a mile to turn around.

8.7 Tanker Availability

Sale Models

In considering this issue we need to again point out that there are two business models we are working with in selling water to China, India, or any buyer including the Middle East. Should our point-of-sale be at source or at market? In other words, do we sell the water at source in Alaska (FOB AK) or once delivered to the markets in China/India (FOB at market)? These different models are based on options that we must consider in any Delivery Contracts. 1. The Market Partner provides his own tankers. 2. We lease the tankers from a third party. 3. A third party company manages all conveyance per Delivery Contract. If selling to a foreign country (not a private buyer) that is in anyway "insecure" we encourage the buyer or market partner to provide their own conveyance, and thereby assume the risk over the 30-year term. Remember, other than the United States all other nations believe "new governments" have the inherent right to renegotiate any contracts signed by the previous government. Or that no "government" can constrain a "new government" in its contractual or other obligations. The US State Department and our Department of Commerce try to hold new governments to existing treaties, but they have little more leverage than we do.

Fortuitous Timing

First off it is important to note that we are entering this market at just the right time as most tankers are being leased and new tankers are being built **at lower prices not seen in five years**. This is a direct result of the previous downturn in the global economy and the reduction in energy consumption. It is also due to the United States becoming energy self-supporting. Not dependent on Middle Eastern oil and tankers. PLEASE keep in mind that energy is a global commodity that can change its price at any time.

Although we have contacted many tanker and marine management companies in our effort to obtain cost comparisons and look at potential tanker leases, we have only been able to work with two tanker brokers/companies. Other companies or consultants demanded significant upfront retainers, or earnest money, or consulting fees just to open an account or they are unwilling to be of any assistance. We were also able to open a direct discussion with **Hyundai Heavy Industries** in South Korea, who builds some of the best VLCC and VMaxx tankers (130 to date). Although we have asked Hyundai for a formal proposal they have declined until we can present them with a freshwater tanker design that they can then cost out. As of this date there is no VLCC or VMaxx class tanker that has been specifically designed for the lift and conveyance of freshwater.

Based on our discussions with tanker experts, we believe the cost of a freshwater tanker will in fact be much less than a standard crude oil tanker because of the elimination of lots of redundancies, spill contingency needs, and firefighting equipment required for the lift and conveyance of crude. It is obviously in our best interest to understand this opportunity for savings when talking with a designer and/or builder. As of this date, given the China virus, we are planning a face to face meeting in Seoul, Korea in the spring of 2021. We also plan an onsite visit to the shipyard in **Ulsan** (Korean pronunciation: [ul.san]). The **Ulsan Metropolitan City**, is South Korea's seventh-largest metropolitan city and the eighth-largest city overall, with a population of over 1.1 million inhabitants. It is located in the south-east of the country, neighboring Busan to the south and facing Gyeongju to the north.

Ulsan is the industrial powerhouse of South Korea, forming the heart of the Ulsan Industrial District. It has the world's largest automobile assembly plant operated by the Hyundai Motor Company;[2] the world's largest shipyard, operated by Hyundai Heavy Industries; and the world's third largest oil refinery, owned by SK Energy. In 2017, Ulsan had a GDP per capita of $65,093, 000 the highest of any region in South Korea.

After that visit we go over to the port city of Incheon to update our understanding and capability of that area.

Then we go to Seoul to speak before the S. Korean World Trade Center and hopefully enter into agreements with one or two proven investors.

Our goal is to secure Letters of Interest from Hyundai Heavy Industries, up to two investors in S. Korea, and a potential partnership in Incheon for bottling etc.

L&R Midland, Inc. (Houston, TX)   http://www.midlandshipping.com/info.php
Malcolm Hounsom, originally from London, has been our best contact. He advises that we may be able to acquire or lease a new VLCC standard design vessel originally designed for crude oil but yet not used; however there appear to be some opportunities to buy vessels already under construction, make the freshwater modifications, and get it delivered in 2021.

BUT we do not want a tanker until we can fill it and have a reason to, so we have some time (upon receipt of a delivery contract) to work through these first tanker acquisition options. We continue to explore these options with Malcolm who has agreed to become our "broker" once we are funded (at a flat fee of $3,000 per month) and obviously a commission once we acquire a tanker. Malcolm is also remarkably familiar with the V Maxx design and is assisting us in the design for water.

Keep in mind that when we are in operation with the first tanker and into a huge market in China, we passionately believe the world will change and there will be many others who will try to compete with us. Our leverage is that we hold 20 sources of exceptional brand power and quality as well as relationships with several other pure freshwater sources around the world we have explored.

Malcolm initially recommended we start the venture with an Aframax class tanker with about 700,000 to 750,000 bbl. in cargo lift capacity. That would be about 83,468 to 89,430 cubic meters or 22,050,000 to 23,625,000 gallons. However, after some calculations for markets in China/India, it became clear we cannot deliver sufficient water to our initial target markets at a cost that is reasonable using an Aframax tanker. Thus, we began to pursue a VLCC class tanker and preferably a new V Maxx design specifically built to freshwater specifications.

He also advises that these existing ships have separate ballast capacity, and to preserve the quality of the freshwater product in ballast tanks may be difficult.

This is one of the design questions we need to raise with our selected designer be it Stena Bulk, Hyundai Heavy, or someone else. Further, it may not be possible to load this vessel fully and get her to her deadweight marks in some market harbors. We are exploring this further with Hyundai Heavy and others as an additional lift option. The draft capability in some harbor areas in the world is over 80 feet so both a standard and a V Maxx design can access our sources and market harbors.

Rates for short term charters are relatively high; however, a ten-year lease could push such lease fees lower as it allows the owner to amortize a substantial part of his capital. If we were to start with a ten to fifteen-year lease based on an actual Delivery Contract, it would likely be in our best interest to purchase the tanker outright, as most have a life of up to 20 years depending on its use for lifting only freshwater because it is less corrosive then crude oil.

Currently tanker owners are looking for **$23,000+ per day** (US dollars) for an Aframax tanker for a 5-year lease/charter, and this is not usually a new vessel, therefore requires conversion to freshwater and all those expenses. A NYK Japan (former partner in World Water SA) the largest shipping company in the world, study found that even if such conversions were done in China, the cost would exceed $1M and take many months and still the remaining life of a tanker of that type is 15+ years. This "lease" cost **does NOT include fuel**, loading or offloading fees, docking fees, etc. We concluded that this was NOT an option given the cost, our return, and volatility of energy costs in comparison to purchasing a new V Maxx designed for freshwater.

Keep in mind it is our clear intent once we have a new V Maxx freshwater tanker designed, we will keep the patent/design. This further enables AQUEOUS International, Inc. to have significant power in this new market and leverage additional incomes.

Fuel consumption/cost for an Aframax tanker is about $23,500 per day. So, combining the lease and fuel cost of this class tanker is at least $46,500+ per day. **This does not allow adequate margin or an attractive price at point of sale.** We refer to this as the *"day cost"* fully loaded. But this does give us an understanding of some cost variables and pushes us again closer to the V Maxx VLCC that is specifically designed for freshwater.

Great Eastern Shipping (India)   http://www.greatship.com/shipping/fleet_shipping.htm

The Senior Advisor to Great Eastern Shipping in India, one of the largest in that country, Mr. P. Sudarsanam, has been working with us in finding potential tankers and outlining the capabilities and limitations of various classes of tankers.

In our discussions with Mr. Sudarsanam, we came to the decision that a VLCC class tanker, built new and specifically to lift freshwater, that the V Maxx was the most cost appropriate strategy. This is when we asked Hyundai Heavy Industries for a proposal based on some initial research that shows them as the fastest, lowest cost builder of such vessels. We now know that China can build VLCC tankers, but their designs would need modification for freshwater, so getting an agreement from Stena Bulk to use their V Maxx design is a solid option if possible and there is concern with the China ship builders. As already mentioned, Hyundai has declined to press forward until we present them with a design as no one has ever built a VLCC class freshwater tanker and Hyundai Heavy is not comfortable presenting any cost estimates without a specific design. They did confirm that such tankers are normally built within 23 months, but that we will need to get in line for space. So, the sooner we can commit to our first newly designed tanker the better. Given our continued research into ship builders, we now believe, depending on the timing, we can get up to three V Maxx VLCC tankers built at a time if that is necessary (weekly deliveries) once we have one or more delivery contracts.

8.8 Distances for Conveyance from Southeast Alaska

The marine distances between SE Alaska and the targeted markets are outlined below:

Market	One-way	Speed	Hours	Days
China, Shanghai	4,490 nm	13.5kts	332.59hrs	13.85 days
India, Mumbai (Bombay)	8,910 nm	13.5kts	660hrs	27.5 days
Dubai/via Spore	9,905 nm	13.5kts	733.70hrs	30.57 days
Dubai/via Panama & Suez	11,319 nm	13.5kts	838.44hrs	34.94 days
Saudi Arabia, Yanbu via Spore	10,946 nm	13.5kts	810.81hrs	33.78 days
Yanbu via Panama & Suez	9,000 nm	13.5kts	666.67hrs	27.78 days

"Spore" is short for Singapore
Canal transit times are NOT in these figures, which have at least 3-day delays and fees as high as $500,000 US.

Figure 35: Conveyance Distances for Southeast Alaska

Costs based on an Aframax class tanker   (comparative example)

The average **FUEL** day-cost for an Aframax tanker is $23,500 per day.

If we **do not** use the Panama and Suez canals due to delays (3 days) and fees ($500,000):

China, OW fuel cost is about $318,550 RT is	$637,100	+2 days	$683,100 in fuel
India, OW fuel cost is about $632,500 RT is	$1,265,000	+2 days	$1,131,000 in fuel
Dubai, OW fuel cost is about $703,110 RT is	$1,406,220	+2 days	$1,452,220 in fuel
Saudi Arabia, OW fuel cost is $776,940 RT is	$1,553,880	+2 days	$1,599,880 in fuel

Figure 36: Aframax Class Tanker Costs

For this class tanker, it will take **at least one day at each end to load/off load**, so we need to add two to four days on each RT in our fuel cost estimate.

An Aframax tanker lifts between 700,000 bbl. – 750,000 bbl.

       83,468.329 cubic meters   to   89,430.353 cubic meters

       22,050,000 gallons   to   23,625,000 gallons

Using an Aframax class tanker under a 5-year lease at **$23,000 per day** and a fuel consumption cost of **$23,500 per day** sets a base cost under a 5-year lease at a minimum of **$46,500 per day** – buy only part of our day cost.

The <u>estimated</u> **cost at point-of-sale** given these conveyance fuel, lease costs, and unit purchase costs along with required time on the sea are:

Market/using an Aframax Tanker	Cost at point of sale
China	$.08 per gallon
India	$.12 per gallon
Dubai	$.15 per gallon
Saudi Arabia	$.16+ per gallon

Figure 37:   Aframax Per unit Costs

Our preferred tanker, the V Maxx, lowers the per unit cost by one to two cents per gallon.

Compared to a new VLCC tanker and given the small size/lift capacity of this class tanker and the high day-cost due to a short-term lease of 5 years, we estimate that these costs at point of sale/at market are likely conservative, but likely high per gallon in the market. There is a lower risk, but you must find a buyer at these prices, and our research suggests that may be difficult.

The efficiencies of cost vs. scale using a new VLCC class tanker, built to VMaxx design standards (eliminates lots of mechanical and contingency requirements for a crude carrier) should allow us to significantly lower these costs at point-of-sale/in market along with less vessel weight.

Costs cannot be precisely estimated: experience will tell
Even when/if we have a proposal from Hyundai Heavy Industries or anyone else, experts advise that we never really know the actual cost per gallon at point of sale/in market until we have an actual tanker, have harvested and delivered the water, and completed a full round trip. A tanker's performance can be very

specifically determined so that its day-cost under various situations of lift, speed, etc. is "actual" not estimated and its unique lift capacity is also "actual."

The average speed used in the above cost estimates for the Aframax tanker was 13.5 knots, but many tankers now travel at 15 to 17 knots, likely cutting a day or more off for every round trip but increasing fuel consumption costs. These **"actual costs"** can change our cost at point-of-sale to our advantage.

> NOTE: These costs at point-of-sale MUST be kept CONFIDENTIAL proprietary information. We NEVER want to disclose our cost at point-of-sale as that could undercut any negotiations with buyers or even some sellers. Loose lips sink companies too.

8.9 Delivering "Drinkable Water"

With 90 million gallons aboard a V Maxx or 84 million gallons onboard a VLCC tanker it will take about 10 – 11 days to enter, fill at source, depart, transit, enter and off load in China. In our discussions on tanker design, we have been advised that it is possible to design an ultraviolet light water treatment system that is the least costly and safest type (no chemicals) of water treatment system for a moving tanker. The UV light will kill all possible biological bugs in our water allowing us to sell the water as "drinkable" at a market point-of-sale. UV light also leaves all the natural minerals in the water unlike filters. This is just insurance but will greatly enhance the value of our water with this protocol. If we sell the water at source (FOB AK) then this is not our concern.

The key will be in the design, safety, and cost against enhanced return at market point-of-sale.

Must Deliver Drinkable Water

If we do not deliver drinkable water at market point-of-sale, the buyer will need to treat the water at their cost. Given the days at sea and the ability to amortize this investment over 30 years, the life of these tankers, this appears to be a cost/operationally viable concept. We will continue to explore this with tanker designers once we have a qualified designer under contract. We continue to be told, "but this hasn't been done before". Yes, we know but in talking with tanker designers, they do not think that is an obstacle once we outline the concept in detail.

8.10   Price Targets at Market/Point-of-Sale

For planning purposes, our all-in cost to China is less than $0.07 per gallon for pristine *natural* Alaskan glacial water. The 2019 local wholesale price of good quality filtered water is $0.25+ per gallon. (to be updated by or next Euromonitor report) The most recent retail cost for high end *'fine'* water in China is over $2.00 a bottle. There is lots of room for elasticity and appropriate margins. We will sell a greatly superior product at a much lower price. When deliveries are at $0.10 or $0.15 per gallon begin, gross margins exceed 100% and ROI is well north of 45%.

Targeting Prices

We have requested retail price sheets from the International Bottled Water Association (IBWA) for both China and India but were advised that such would need to be purchased. Once acquired, these proprietary price sheets, at well over $1,000 each, will give us a better idea of target pricing for these markets, however, we will continue to focus on the mid to high-end market at least initially to set a mark for our image in the industry and in each market – solely based on the nature of our pure *natural* Alaskan glacial water.

Euromonitor is now providing some new intel on each of our first ten target markets in China. This will help adjust our margins as necessary, but the latest intel back from our lawyer in China is that demand continues to accelerate for bottled water and what we expected the consumer is getting educated well on the problems of water contamination.

As mentioned, we have also opened a relationship with Grail Research and have a proposal from them to do extremely specific targeted market studies in all likely markets starting with the ten in China. This is on hold pending ongoing negotiations with Euromonitor International (our preference) who can also provide this service.

Then we got COVID and all the delays attached.

Later, with sales/delivery contracts in hand we can adjust our margin to allow lower end sales in some of these markets if that makes economic sense.

The price of a filtered 5-gallon (just under 19 liters) bottle of delivered water in India, we are told, is $2.00 - $5.00 US. The five-gallon bottle, either round or square, is considered the **industry standard globally** for commercially delivered water. Thus, we know that even at our top price, they are realizing between a 20% and 30% return on their investment, we present a uniquely higher quality, higher brand marketing potential product that is very competitive in the India and Chinese markets.

9. STRATEGY

9.1 Business Models (two)

In discussing delivery contracts with buyers in China initially and then India, these two models basically attempt to address the question of who takes the conveyance risk.

If we sign a Delivery Contract with a buyer in China for a minimum of 30 years to allow us to fully amortize our capital investments (mainly site development/O&M and conveyance), what assurances do we have that the buyer will live up to that 30-year contract provided we are performing as agreed? In foreign nations this can be a challenge. So, to address it we have basically two approaches:

---

1. We provide the conveyance from Alaska to market
    a. We purchase the vessels and operate them, or
    b. We contract with a third party to purchase the vessels and operate them
2. The buyer or market-partner provides their own conveyance
    a. They purchase their own vessels and operate them, or
    b. They contract with a third party to purchase the vessels and operate them

---

Figure 38:   Two Business Approaches

We provide Conveyance

If we provide the conveyance, we take the risk. If we lease a tanker or contract with a carrier, we also have this risk, but not as much. Remember, one V Maxx VLCC tanker costs about $150 million. To fully exploit just our water sources in Alaska, we would need a minimum of up to nine tankers - $1.350 billion per source in conveyance risk. Rather a steep barrier to entry for competitors – we like it. So, contract stability over 30 years is essential. Frankly, we like 99-year delivery contracts with 5 year adjustments even better.

Buyer Provides Conveyance

If the buyer provides the conveyance, especially to China who builds VLCC class tankers, we shift the risk to the buyer and thereby apply stronger leverage to ensure the 30-year Delivery Contract. China has the investment in the tankers (what we Americans call 'skin in the game'), offloading systems, jobs, production, economies, etc. so we also have some market leverage in the government.

Avoid Canadian Mistake

These are options we MUST keep on the table as we work with prospective buyers in any foreign nation. In the early 1990s twelve companies had signed contracts to purchase water from sources in British Columbia, Canada. Then the provincial government changed, and the new Green Party administration refused to even recognize those contracts. This resulted in the largest private trade debt, yet unresolved as far as we know, settlement between the United States and any other nation. And you never heard of it did you.

Contracts are not the same as in America

Similarly, World Water, SA found themselves, after a few years of operation and significant investment in WTB w/tows, with a new government in Turkey where they had been purchasing bulk water for conveyance to Cyprus under a 20-year agreement. The new government of Turkey decided that the negotiated water sale price was too low and arbitrarily changed it. Cyprus, the buyer, refused to pay the difference and conveyance stopped. When World Water, SA asked the US State and Commerce Departments what could be done, they said basically nothing. Contracts are not the same in any other nation as they are in America. This is why we stopped looking at other sources then SE Alaska – and very glad we did.

So, that is why we have two business models. Who takes the risk on conveyance in a 30-year delivery contract?

9.2 Short Term (5 to 7 years)

The purpose and function of this **Business Plan** is to outline the initial viability of this venture. We may refer to this as "Phase A" or "Startup" precisely because it is only the first of what will likely be several phases in the development of a **global** bulk water acquisition and holding company focused on **global** freshwater market opportunities. This phase only looks at establishing the company globally through the acquisition, export, and sale of an extraordinary high quality and therefore high value "secure" pure *natural* Alaskan glacial water from 20 sources in southeast Alaska.

Opportunities

But these are critical first steps in major markets that will likely open to us an even wider array of opportunities once we are established and have solid market and even international investment partners. By creating our image/presence through the import of exceedingly high quality pure *natural* Alaskan glacial water we establish a very marketable image in the two [China (10) and India (2)] most immediate and demanding freshwater markets on earth. Growing from this point of entry is a significant advantage to our longer-term objectives of providing freshwater at low cost to a range of markets in need anywhere on the planet.

Obviously the first order of business is to create a sense of comfort in Alaska with this company and its ability to lift large quantities of high-quality *natural* Alaskan glacial water for global markets and do it with ecological and cultural sensitivity. Securing the ownership of the water is the critical first step along with the export license from the state, both essential. To do this we need additional capital. We have a potential investor in Alaska that is capable and has knowledge and experience in water. He chose to wait and see if we can attract other investors. The Chinese virus shut our market development efforts down due to travel restrictions. We are now back in the game and are getting calls from both American and foreign investment groups.

Three-year Requirements

We have said for some time, and it is again proven with the 2019 numbers we gathered, that we require about **$6M for three years** to make this work. Keep in mind that many (60%) of these costs are one-time per source, or water 'take point' related and as long as we take water the sources are ours for eternity. Ownership of sources provides leverage and investor security. Included are operational, marketing and personnel/staff/consultant costs, but they are nominal against the cash flow of the company once a delivery contract is signed and we begin full site development, order a tanker if that is the buyers choice, and start drip marketing and then sales. One source, servicing one buyer in China, with water selling at $0.10 to $.15 per gallon monthly average income ranges from $2.9 million at $0.10/gal to $7.0mm at $0.15 per gallon – **every month** for 30 years of a delivery contract minimum. Again, that is one source, with one tanker, with one market, with only one lift a month. But we can make up to 42+ lifts a year from just one source with our ownership of that amount of water, using some incremental infrastructure and only adding three more tankers that ensure weekly deliveries to market. The multiples are then enormous.

We have told the state that we plan on lifting our first water within three years given the time required to secure a delivery contract and build the loading facility and procure lease/acquire our first tanker. It will take us no less than 24 months to do our first lift given the 23 months it takes to build at least one freshwater V Maxx VLCC class tanker from S. Korea, and complete sea trials.

Within the first three years of this venture, it is our intention to create a serious global interest in this company, its secure and unique water sources in Alaska and many other regions of the world, the quality of our product, and the ability to apply new and futuristic technologies, such as energy cost reductions for tankers that we have explored.

Plenty of Interest
Freshwater needs in a wide range of markets/economies across the planet are thirsty. We know already that there has been interest from South Korean beverage companies based on inquiries from potential buyers for water export permits from Alaska. They are unaware of the access and

scale limitations on conveyance there. We also are advised that there is serious interest by several Middle Eastern countries to include Dubai, Saudi Arabia, Jordan, etc. on the import of high-quality drinking water to replace or mix with desal. WE also have ongoing contacts with investors in Canada (bottled water companies) and Los Angeles, California. In each case a Delivery Contract **must** be secured <u>before</u> we can prove actual markets at cost/price. **Delivery Contracts for bulk water are bankable paper** and obviously essential prior to any significant outlay of capital.

To accomplish this, a minimum of $6 million is appropriately needed to "establish" this company globally over the first three years. Additional capital will be necessary for the acquisition and/or lease of tankers and other conveyance related needs but those are more market and source specific and will be outlined in the next Phase of this venture. We have also been approached by the AIDEA (*Alaska* Industrial Development and Export Authority) a well-established quasi-government entity to provide funds for initial infrastructure (take point, pipeline, pumps, etc.). And I didn't even ask.

We passionately believe that with one source, one tanker and one 30-year delivery contract we can then in year three start to cover the initial $6M and begin building our own financial reserves.

Market Analyses
For example, we have several specific market analyses to acquire, and vetted bulk water brokers to interview in China and later India provided by Euromonitor. We have lots of pre-shelf marketing we need to strategically initiate with a 90-day drip and at least a 40-day push before product introduction in first markets. These will all be led by our marketing expert on staff. We need a loading facility built in at least one source, although we would prefer two of our 20 sources to load the tankers. We have secured the services of a major engineering company (PND) with decades of Alaskan experience to assist us in looking at the water take and delivery options and costs here (see 5-year budget). It was the preference of this venture to construct this loading facility totally with private funds so that it remains under the control of this venture,

but the Alaska Industrial Export and Development Authority (AIEDA) has met with us and offered to finance the first two infrastructures to get us started – but we need a delivery contract first. Yes, we will secure a Letter of Intent in our next trip to S. Korea, but we are never 'secure' with just that. They have suggested that this be a joint public/private venture and that if a private party makes this investment, we could offer to have them paid back in water i.e. CA. Again, more flexibility to our advantage. This approach will thus allow this venture to immediately recover its infrastructure investments within the first year of lift and full-scale operations – again from one source, with one tanker, and one market. Then we grow.

Similarly, in the acquisition of the water sources we already know for AQUEOUS International in Southeast Alaska, Albania, Russia, Canada, etc. - additional capital will be required to "purchase" the water rights or licenses and ensure their "security". We will not support any investment that is not as secure as water rights are in Alaska. In both cases additional capital investment may be needed, either in partnership with these sources and/or market partners with exclusive rights to this water that would enable source development allowing the loading of appropriately sized tankers for each market. Our immediate focus will be water sources for China from SE Alaska and then sources in the Atlantic or Mediterranean (we have investigated many already) to serve Middle East markets if they can be secure.

Expansion of Bulk Water Buyers

The ongoing expansion of bulk water buyers in such markets as Canada, who call us often, then with an amendment to the Jones Act, Southern California between San Francisco and San Diego offer **significant** opportunities if, **and only if**, the imported water is understood to be of superior high quality, high end branding, and predominately for packaged drinking. We even have an interest from a beer brewery in LA – because the water they receive from LAW&P requires high filtration costs and they like the AK Glacier brand.

Due to the historical level, and institutionalization of federal, state, and often local government subsidization of municipal waters, imported water cannot compete at the subsidized price. If these subsidies were ever lifted and west coast consumers then forced to pay the actual cost of their delivered water, imported bulk water would likely become extremely competitive in these markets for municipal buyers applying our municipal 'mix' strategy with first preference any existing desal plants. No serious studies that we can find have been done on this question of quantifying the actual value/cost of these federal, state, and local subsidies. But we do know that the 'actual cost' of water and of these subsidies to governments are causing fiscal structural concerns, especially in California one of a dozen states that are already financially bankrupt with no way to crawl back. Their income bases are all leaving and that has accelerated with the inane shutdowns.

9.3 Long Term 50-year Plan

It is the intention of this venture to establish itself as "THE" global company to provide and in fact at some point "control" a majority of new privately owned/licensed freshwater sources and be known for its development of technologies to explore other sources to immediate and emerging markets across the planet. This will be accomplished with the accumulation of significant water assets, Delivery Contracts in strategically targeted high profile regional and national markets and applying the solution or technologies that make the best business sense for each specific market and this company. Our biggest challenge will always be 'ownership' verse a government permit. The only way to fix this in our experience is with an in-market source partner who has not only the economic strength but also the political strength to overcome this limitation.

Long Term Generational Strategy

The acquisition of new technologies in the development of and the acquisition of other such technologies or intellectual properties will significantly enhance our portfolio and return on investment of this global company. This is an important component of this company's long-term generational strategy so that it will remain at the forefront of this emerging global commodity market – with a natural renewable resource unlike nonrenewable oil. There are in fact only a few high-quality natural water sources available at tidewater. We have looked at multiple dozens as we have helped nations around the world address their water shortages and their growing concern with desal water quality to human health and source security (very exposed to terrorists).

Separate Contracts for Separate Projects

Based on the experience of the members of this venture, each "project" for example one in Albania, should be organized and constructed as a legally separate venture/entity with appropriate source and market partners under the control (retaining the largest percentage of shares) of AQUEOUS International as the global development, water source, and holding company. This business construct enables market, application, production, and marketing control by this venture - but allows the local partner(s) at source or market to appropriately be the local face and strength of each venture. Each venture must be its own profit center with an appropriate percentage of profit going to the holding company that holds the water rights if applicable.

We project, with our entry into this market, that within ten to fifteen years water will, in most areas of the world, become an actual globally traded commodity. It has already started in California and is moving into other western states. As such, the financial opportunities that will open to this venture are then extraordinary. Global market trading and futures in water will become as common as oil and gas trading and futures. To have a global holding company with several strategically located and significant (quality/amount) water sources, unique production, or conveyance systems in place at that time, and in control of unique technologies that can create "new water" such as the freshwater sea vents, will enable the investors/partners in this venture to be at the right place at the right time with the right "control" over global markets – and make exceptionally good money with little ongoing effort
Cost per Cubic Meter or US Gallon by Market

Cost per Cubic Meter or US Gallon by Market

Generally, the comparative cost of desalination, if all costs are included (and they rarely are) including all subsidies apparent or hidden but based on state-of-the art Reverse Osmosis (RO) single or multistage desalination systems is $1.00 to $1.50 per gallon. In almost all cases this will be the competitive cost, but we need to continue to point out that our pure natural Alaskan glacial waters are not desalinated, nor filtered, nor 'manufactured' waters with a significant loss of natural minerals, but fresh natural glacial water from rare arctic sources in SE Alaska. When retailed in bottles or in other beverages, our per-unit value is significantly higher, and buyers intuitively expect this, than the value of municipal desalted water. Remember, we now have scientific proof that the consumption of desalinated/demineralized water is harmful to embryos and young children (proven in Russia in 1965). Doctors now instruct pregnant women NOT to drink desalinated water and, in some areas, (even in America), and not to drink any municipal water because of its chemicals, and other contents such as pharma. This causes a strong emotional consumer concern, especially with young women and new families concerned with what people drink for the rest of their lives.

Some of the key numbers:

- ✓ Our cost of water at our Alaskan point of lift is $0.0355 per gallon.
- ✓ Our **all-in** cost of water acquisition & conveyance at China is $0.07 per gallon.
- ✓ The 2020 wholesale price of filtered freshwater in China was at $0.28 per gallon.
- ✓ the retail price, of water **on-shelf** in China per container has an enormous range we will have updated once we obtain the latest report from Euromonitor.
- ✓ There is a sufficient pricing level to support extremely healthy down-chain margins at our target bulk, wholesale, or retail pricing

Figure 39: Key Numbers

Price by Market

- *Price must be greater than cost, plus an adequate profit margin to ensure liquidity. Price is what consumers are willing to pay after marketing*

Price Elasticity
Price is market driven, not cost driven and **must include the price elasticity** of delivered freshwater **in each market**. For example, there are markets now paying $10.00 US per cubic meter for delivered water in the Mediterranean. With known water sources and appropriate proven

conveyance, water can be delivered to these markets for far less than half of what these markets are now paying. This is based on work AQUEOUS International did for the government of Italy a few years ago. The biggest obstacle there was, and remains, organized crime who has totally controlled this business for generations. Wish the Italian government had told us that before we came face to face with 'the Boss'.

It makes much better business sense to offer the water at just less than the current price of a *comparable* product in each market as opposed to cost plus an arbitrary margin.

Even with a 30% margin in these markets, we could have saved the Italian government tens of millions in US dollars a year. That is why we need to know and clearly understand the elasticity of these unique water wholesale and retail markets based on their intended end use, so that "price" is appropriate to that specific market, not just cost-plus margin.

Competitive Prices for Pure Water
We now know that water wholesalers are selling bulk so called "filtered" water in China for bottled beverages at about $.25 per gallon. This is locally filtered/treated local water that is most likely contaminated, some even worse than required American drinking water standards, has lost important minerals in filtration, and still retains chemicals, pharmaceuticals, and in some cases even fecal coli as reported in the latest series of bottled water contamination reports mentioned in this plan. If these numbers are accurate and comparable, our target price for our pristine *natural* Alaskan glacial water, which is of significantly higher quality and brand marketability, should be competitive even at $.20 per gallon wholesale.

10. DELIVERY CONTRACTS

Delivery contracts are the essential commercial paper that makes this venture work and enables essential credibility and financing before sellers of water sources. Delivery Contracts are also a "bankable" instrument for project financing in most global markets. What percentage of financing depends on the economies within which a specific project will function? Certainly, in the United States today, this will be a much lower percentage then what it would have been in the early 2000's. What it might be in China or India is another question. One we cannot answer this date but will update soon with the acquisition of proprietary reports we have contracted for our first ten markets from Euromonitor.

Function of Delivery Contract

The problem is that most people do not know or understand a real Delivery Contract. In summary, what it does - is say in writing what is 'contractually' agreed to, but the performance is almost exclusively on the back of the seller and in our case the water provider. If we are selling the water at point-of-take (in Alaska) our financial exposure is extremely low. If we are selling the water at market (China), and therefore providing the conveyance via ownership or lease agreement, our exposure is dramatically higher.

Simply, the delivery contract outlines that:

- A specific amount of water from a specific source is quality tested and delivered to a specific point of sale based on
    - An agreed starting date or first day of delivery
    - An agreed schedule of delivery (always allow some room for delays)
    - An agreed water quality (tested at the point of take and point of delivery)
    - An agreed delivery quantity (cubic meters/gallons)
    - An agreed period – must be a minimum of 30 years if we are responsible for conveyance. Otherwise, we should be able to amortize our capital investment at point-of-take in 3 years or less. We like functioning without debt.

Figure 40: Delivery Contract Stipulation

If the above stipulations are met, the buyer then agrees
to pay for this water based on:

- An agreed to price per unit (gallons, cubic meters)
  - Such price <u>must</u> be negotiable/adjustable annually based mainly on the cost of energy (a critical flaw in former ventures as energy costs are so volatile)
  - A payment schedule generally using advanced wire deposits released at time of departure from source or delivery (ability to immediately confirm payment is essential upon completion of load if that is our point of sale, or payment upon initiation of offload if that is our point of sale) This is all negotiable.
  - An allowance for other costs to increase either on a pre-agreed schedule (5 years) or as may be negotiated. Some buyers may insist on percentage caps but given the volatility of energy this is not advisable.

Figure 41: Delivery Contract Buyer Role

Most agreements allow a five-year reopener to allow for some negotiated adjustments based on schedule/demand, and or cost/price – almost always the volatile changes are in energy costs. But it is essential that such reopeners do not compromise the longer term of a contract. There are very experienced global firms who can develop cost/price projections based on water sources and each market. This type of service should be used no less than every 3 to 5 years to ensure our actual costs are covered and we are able to retain our margins.

Minimum 10-year Contracts
Delivery Contracts <u>must</u> be at least ten-year contracts to allow adequate amortization and/or depreciation of infrastructure to work. Most are for 20 to 30 years, but some Delivery Contracts can be up to 99 years, especially if you are dealing with a municipal water supply critically dependent on this import – some daily. Economies fail rather quickly without a secure clean water supply upon which that economy is based.

Once you have a Delivery Contract you have a bankable paper. The contract is then provided to any major financing group and you can "finance" a negotiable percentage of that contract based on some agreed margin and payment schedule. We prefer to do this within the United States, but we know some will want to do it in their market, so we must verify prior performance and have someone watching this regularly.

10.1 Delivery Contracts are essential.

Letters of Interest are good to have and we will collect them from prospective buyers, but they are **not a contract**. Memorandums of Understanding or Agreement are also offered but are again not bankable on our end and not generally accepted for signature by financial firms. Often, in the water business, companies present Letters of Interest to persuade someone of a serious interest. In our experience, these are often not worth the paper on which they are signed because there is no real or implied "contract" and no skin $ from the buyer in the game from their side. It is just 'feel good' politics. *Unless there is a real banked deposit towards a deal,* which can be an option in the form of a down payment in escrow.

Getting a Delivery Contract allows you to show a real bankable document from a committed buyer with at least some specific set of terms to an investor or bank. These Delivery Contracts can only be changed with the written agreement of **both parties**, and they can set any start date or delivery schedule that is factually viable from an engineering and financial base.

Anytime we secure a signed delivery contract from a foreign market, we will provide a copy of it to the local American ambassador, the US Department of Commerce, the US State Department, and the US Trade office. We may be asked for a copy by the CIA as well – been there done that. We have learned that all of these are in our best interest.

Some new buyers get nervous when a Delivery Contract is presented thinking that somehow, they are locked in. Yes, a contract is a contract, but the "performance" is totally based on the behavior of the seller, and not the buyer unless there are buyer responsibilities that are agreed to such as their own conveyance. If a potential buyer is not willing to sign a Delivery Contract, then the ability/commitment of that buyer is in real question and most serious people in the business understand that. Once the price, amount of water, and delivery schedule is agreed to – with agreed margins – we MUST get some funds from the buyer to 'hold' that agreement until signed.

NOTE: It is possible that some market buyers will ask for a percentage of our water source that has provided 'their' water. We will not agree to this as it may compromise our licenses, permits, contracts and sensitive political persons who stand with us. This may happen after a decade of success from a source, but we MUST ensure this possible shared international 'ownership' does not compromise us politically by opening other sources to foreign interests. We are VERY caucus of the Chinese gray zone strategy that has compromised many countries only to the advantage of the CCP and not the people or economies where the sources are located.

In our presentations to state or federal agencies, we will need signed Delivery Contracts if we are to have the highest level of credibility and secure any additional permissions or assistance, such as 'extensions' in permitting permission processes. WE DO NOT PROVIDE ANY STATE OR LOCAL AGENCY A COPY OF SUCH CONTRACTS, we can *show them* to the official paper, but we do not leave them a copy as they then become public property, and anyone can access them without any restrictions. We are currently setting up the agreements with state agencies that our contracts are proprietary and not for public release. This is within the law, but few know it. We have 'proven' that our venture is viable and will just take a little more time for whatever reason – usually weather especially in SE Alaska. The reason this is a problem is that there have been lots of scams presented to government agencies that have caused political problems. We stay awfully close and on top of this issue with meetings all the way up to the Governor and AK U.S. Senators and Congressman whom we know well.

Alaska has had bad experience with Letters of Interest, MOUs, MOA, etc. as developers often claim to have contracts in hand or pending, but never produce them for required public confirmation. These 'promises' are leverage for investments, and then it falls apart and the investor is left with an empty wallet as the scammers pack up and move to a new opportunity. We have been consulted often by the U.S. Security Exchange Commission, many Wall Street Firms, and others on just these problems.

Private proprietary documents can be withheld so long as the state or federal agency has **seen** them and knows and can speak with confidence that they are real. You do not have to fully disclose price/cost etc., only that you have a contract for a period (30 years is our base) and a start date. This can be done by affidavit as well.

11. COMPETITION

11.1 Direct Competition

First, it is essential to understand that we are NOT selling just any water for any purpose. We sell extraordinarily high quality pristine *natural* Alaskan glacial water (internationally graded as "fine waters") for bottled water/beverage markets, industry, and hopefully municipal mix. If we have a broker who does not want to disclose their buyers, so long as we have a Delivery Contract with the broker and have verified its veracity with some financial escrow, we stand ready to perform.

In the minds of some, Alaskan water will "need to compete" in the bulk water marketplace with desalination, stormwater reclamation (some from sewage), redistribution, ground-water management, land-based delivery systems, conservation, expanded storage, water marketing, and new source development. Except for desalting and bulk imports, **any new source development is totally dependent on making better use of <u>existing</u> water in market resources. <u>All are also, to a significant extent, directly dependent on local weather</u>** - *except* desalting and bulk water imports. This includes in most cases even groundwater.

Efforts will continue in many markets to enhance existing sources, but these techniques have become increasingly constrained by economic, political, and environmental forces as population and pollution expand exponentially and national and state efforts to dedicate more water to fish, wildlife, and recreation increase especially in markets like California, Oregon, and Washington state. We have paid very close attention to these obstacles. Really only two prospects offer significant potential for providing acceptable secure *"**new water**"* to a market area allowing sustained growth: 1) bulk water imports, and 2) more efficient, environmentally acceptable, and lower cost desalination – but only if they discard the adverse health effects of demineralized water and the growing concern with bine discharge that is already causing serious cost increases and ecologic trauma in a growing list of living organisms.

This Business Plan's water is initially targeted toward mid to high-end bottled water or beverages at the wholesale/retail levels allowing a far higher price per gallon at sale than anything you can realize selling to any public water system. This allows for a much more attractive profit margin and international brand image. It also allows our brand to be seen, gain consumer confidence, and be attractive to other potential buyers such as industry (there is serious industrial interest especially in China) and municipal mix. It is **_essential_** that our market's 'top-of-mind' response to our pristine Alaskan *natural* glacial water be the expectation of any prospective buyer.

So, the comparability of bulk water for municipal systems vs. bulk water of high quality for bottling **is not correct**. They are different markets and different economies of scale in almost every case. The cost price margins of delivering bulk water for bottling within the elasticity of the bottled water markets in each market region are significantly wider.

This venture may see some potential competitors just before our product presentations in China. But based on pure natural Alaskan glacial water we already know of high interest by bottlers in Canada, South Korea, and China. Understanding who they are and what they claim to have in Delivery Contracts, rather than just Letters of Interest or MOUs, is essential to our success in arguing for the opportunity to secure Alaskan or other sources.

We have followed closely the "pronouncements" of various companies as they continue to make claims in their supposed use of the Sitka/Alaska source for example, which appear to be timed and structured to impress investors and influence stock sales more than any actual water reality. **To date they have not sold or moved any water**. There are lots of "claims" and initiatives, but the bottom line is:

- They cannot get a tanker of sufficient scale into the Sitka source. (they physically cannot)
- Do they have actual signed verified Delivery Contracts? No!
- Can Sitka or the US Departments of Commerce and/or State

(the US ambassador in that market) independently verify such contracts? No!
- What is their cost/price points per market compared to what we know? None!
- Do they have the financial and operational capacity to perform? No, they are betting on a buyer who doesn't understand the source limitations. That is why the SEC calls us often.
- Do they have contracts to lease/purchase tankers? No!
- What has been their history of performance? None!

As we have stated to Wall Street, international investors and several federal agencies, et al when we receive calls asking for our opinion of the Sitka source. We answer that *'it is a scam'.* We hate to do that, but that is what it is and will continue to be until the tanker access problem is resolved if that is even possible – certainly not based on our onsite visits and those of our taker captain.

11.2 South Korea

The bottled water industry in South Korea is a potential competitor in our targeted markets in China and possibly India. As we pointed out, beverage companies from South Korea have inquired from time to time about the so-called bulk water export contract in Sitka, but after our discussions have never pursued it. Several companies in Korea have decades of exposure to the waters in Alaska and are currently involved in ventures with water bottlers in Anchorage whom we are personally familiar with, but these are nothing compared to what we offer. Anyone in water in Alaska is known to us because of our reputation and knowledge.

One company in South Korea was just announcing (2019) a new serious of product lines using Alaskan water for South Korea such as, **Baby Water** and **Senior Water**. Although it is all "just marketing", they are showing a very innovative approach to building market share in South Korea and may find it attractive to compete with us in our bulk water export venture. Remember, this is all based on the export of water from Sitka. We have lived and worked in South Korea for over 2 years and have worked with the bottling companies in Alaska with long ties to this market. We are now approaching them as potential investors and partners for the China markets likely late April or early May 2024. They have previously expressed an interest in exporting water from us, so it is important that we keep a sharp eye on their efforts and attempt to be more impressive and responsive to their interests than would be competitors.

With the economic and political pressures between the United States and China over the past three years, our approach in South Korea is now different as we had outlined earlier. The growing trade conflicts between America and China, added to the already known and serious concerns with corruption in markets in the ten largest cities there (our target markets) and the influence of the Chinese Communist Party (CCP) in the industry there. It makes solid sense for us to explore a partnership in South Korea. Chairman Davidge lived in South Korea for over two years in the '60s and has visited it often. This option is to deliver Alaskan pristine natural glacial water to Inchon harbor, which is only 400 miles cross water to China, and have it bottled in South Korea in one gallon or less sizes for export to the ten largest Chinese cities. This can be done via container vessels and the individual containers delivered to the wholesalers in each of these cities.

We are also interested as stated earlier, in the application of WTBs for bulk shipments from S. Korea to Chinese. Our research indicates it is possible to use 35,000 cubic meter to 50,000 cubic meter WTBs from Inchon to manufacturers or municipalities in China for those who have expressed interest in importing large quantities of the Alaskan waters.

The business environment in South Korea is exceptionally good and Inchon is an excellent commercial harbor area that includes several islands. We are now actively involved in exploring this option through our relationship with the World Trade Center as it may solve some of the problems we see in China, but at the same time have our products in these ten targeted cities.

11.3 Other sources of water for global markets

Water is not always 'just water'. The *natural* pristine glacial water harvested from our Alaskan sources is of extremely high quality and a rather unique water destined for bottled water or beverages. It is generally regarded as a 'fine' water which is the top of the international market most often presented in extremely attractive and expensive glass bottles. But unlike other fine waters around the world, we have an extraordinary amount of it. The electrochemistry of this water is quite unique and one of the reasons it is graded as a 'fine' water by the International Bottled Water Association. With glaciers at the age of 10,000 years to 32,000 years it is incredibly special – the older is rare.

With respect to our interest in selling bulk water to municipal markets for a mix strategy (thereby lowering local source contamination), in most cases, municipal markets have depleted any close freshwater source(s) (within even 500 miles) and are now forced to rely on desalination or wastewater purification/recycling to augment their principle natural sources. In almost all cases, where there are bulk water sources within reasonable proximity to the sea and these markets, desalination or water purification is a far higher cost (economically and environmentally) per cubic meter then the cost for bulk water delivered - when all costs are considered, including government subsidies and environmental costs. Additionally, desalination and water purification are more immediately and marginally impacted by the volatility of energy costs then bulk conveyance due to the amount of energy they consume per cubic meter/gallon of delivered water unless they are co-located with a large energy producer such as the case in most Middle East markets that 'give away' natural gas for the energy to desalinate seawater. Further, in almost all cases, municipal water is heavily subsidized and not in any way comparable in real/actual cost to the quality or uniqueness of the water from Alaska. This is where other sources may be of value, but again our main concern is always source security. If you cannot ensure source security for at least a decade – better 30 years, do not waste a dime on it.

Once this venture is underway, we will begin securing other sources of water for these and future markets IF we can ensure source security and obtain some financial commitment from buyers. For example, we know of other high-quality sources in southeastern Alaska that are attractive but more expensive to develop. There are also others that are not glacial. These could be immediately acquired as soon as we have market sale confidence. We also know of extraordinary high-quality waters in Albania that can easily serve Corfu, Italy, Greece, and several Mediterranean and Middle Eastern markets. The cost at market of these water sources is far less than if we import bulk water from Alaska for these regional markets, but they do not have the same bottled/beverage market cache as the *natural* glacial water from Alaska. The Albania sources can however be marketed for mixed beverages and municipal waters because of their extraordinary purity from huge artesian sources.

Similarly, there are developed high quality groundwater sources in southeastern Russian islands that could significantly lower the cost of bulk water imports to especially northern China and even India. These sources have been developed and are ready for bulk export – the problem again is source security, Russian politics, etc. – so we continue to seek a source partner. We have sent messages to the owners of these exceptional freshwater wells about reopening discussions for the bulk sale of their water. Again, the challenge is **source security** over the length of any Delivery Contract and the often-requested under-table payments. (don't ask) We do think these sources are worth pursuing if we can find a competent and responsible Russian partner we can trust.

The idea is that once we are underway with a strong dependable cash flow, we must start the exploration and development of alternative sources that can open new markets or lower the cost per cubic meter for expanding the markets we start with. For example, we can deliver high end water to China for bottling and then also provide lower end, but still pure *natural* water for municipal mix from other closer sources. We believe, based on our experience, that we can acquire at least 93.6 billion gallons of exportable high-quality water from Alaska within two years with adequate financing via delivery contracts. Again, that is 4.6 billion **gallons per site/source per year**. Once acquired we can hold them without any development for at least five possibly seven years before they might convert back to the state under the "use it or lose it law" now almost globally universal. However, with the political and business relationships we already have, so long as we are showing progress and likely contracts, we can keep pushing that envelop.

11.4 The Costs of Options

As we pointed out earlier, the comparable cost of desalination and water purification are between $1.00 and $1.50 per cubic meter. Israel is reportedly now desalinating water at a cost of US$0.53 per cubic meter but that is **with large energy cogeneration subsidies**. Singapore claims to be desalinating water for US$0.49 per cubic meter again **with cogeneration subsidies**. These are both RO plants using the latest technologies. Keep in mind the longer a desal plant operates and discharges millions of gallons of brine in the area, the sooner that comes back into the plant and causes significant additional cost and environmental concerns.

But, at what level of subsidy are these costs underwritten? This enormous cost offset is not reported by the desalination industry, so we do not know. Working on it.

Prior to any face to face meetings with Singapore, we hope to have this intel. Are they receiving free energy, which is very often the case especially in the Middle East? Or are they co-located with, for example, nuclear energy generators or other co-location strategies, and thereby producing manufactured demineralized water with that enormous energy subsidy? If you took away the cogeneration or other undisclosed "public subsidies", the cost per cubic meter would go up dramatically. The fact is the 'actual' cost is that, but consumers are unaware of these subsidies. The key is to know what those hidden costs/subsidies are, how stable/secure they are, and how to present them as a part of the 'real cost' of this water production system, as you attempt to show them alternatives. For us the secret is marketing. And remember, here we are comparing manufactured demineralized water that we now know is not healthy for babies because it does not provide the essential minerals necessary for brain development that is 'organically' pristine *natural* Alaskan glacial water. They are not comparable. Use manufactured water for anything but mammal ingestion, and it makes sense. Use the packaged *natural* water from Alaska and you have the benefits of both worlds.

First, we will help inform our markets that they should not compare desalinated water in bottles with *natural* glacial water from Alaska in bottles. If the cost of delivery of bulk water from Alaska is within the bottled industry's ability to realize a very reasonable profit, and it is recognized by the buyer that <u>the retail value and human health value of the water</u> from Alaska, due to its uniqueness is much higher, we remain extremely competitive as a beverage or base/stock import.

12 DEVELOPMENT PLAN

12.1   Key Elements of the Project – and anticipating questions

Globally the width or radius of the pipe and the number of such pipes at tidewater usable for freshwater determines the capacity of any loading or off-loading facility and their relative costs.

On-shore vs in-water WTB storage is a strong option, but not one we at AQUEOUS International assume on its own. In-market storage should always be the responsibility of our market partner or buyer or at worst the joint responsibility of seller and buyer. If a buyer is not capable of providing such storage, we should consider working with them in the design and development of what they need to ensure our off-load speed and adequate storage, but this must be a joint responsibility in foreign markets to protect our long-term interests.

Some buyers may initially want to have the tankers offloaded into rail cars or tanker trucks. This is an awfully slow process and will significantly increase the cost of offloading by as much as 3+ days doubling the off-load time required as the tanker remains at harbor at cost. If a buyer wishes to use this method, which we do not recommend, we need to add significant additional charges for remaining multiple days at harbor. If we make these additional costs to market high enough, they will develop at tidewater storage or manufacturing. In the terms of any Delivery Contract this is a key issue that MUST be identified and resolved upfront.

The greatest efficiency of a tanker is under lift at full speed, not sitting at a dock loading or offloading. The objective is to get it in and out as quickly as possible, so you can maximize your day costs "at sea". The rule of thumb is 30 hours at dock. But in Alaska we will most likely require 3-5 days for receiving water and two days to come in and leave.

As we have pointed out, in the western U.S., bulk water is sold and purchased based on per thousand gallons or acre foot (325,850 gallons). We will be paying the state a 'conservation fee' of $5.00 per acre foot or $1,381.00 per lift. In the rest of the world, water is sold and purchased in cubic meters. In our markets we need to sell it per gallon or its liter (3.79) conversion because that is the product with margin we need to know and understand.

- 90,000,000 gal per lift, that's 276.2-acre feet per lift, at $5.00 per acre foot = $1,381.00 per lift paid to the state (*this will likely change as we evidence success*)
- Obviously the more we lift per site per year the lower the O&M cost per lift, but it still takes about a day to come in, a minimum of 30 hours to load, a day to depart, so we average, with room margins, 7 days per tanker loaded.
- With the cost the water, site amortization, and conveyance to China the all-inclusive cost is about $0.07 per gallon. We believe we have significant flexibility in margin at point of sale.
- Compared to the on-shelf retail price for 'fine' waters the market partner will likely still realize at least a 40% profit per unit.

12.2   Loading/Off Loading limitations

The limiting factor in loading bulk water to and from a tanker is the radius of the inside of the water pipe and the number of pipes it comes out of. Obviously, distance from source is limiting and pumps can only push so much water through any given width and number of pipes. We solve this at source with the appropriate engineered systems, but at market, there is often a rub.

Most ports have water lines at about 12 inches in diameter principally for firefighting. There are some exceptions, but they are generally not good for freshwater. We need a minimum of 18-inch pipes to realize any load/offload efficiency. This can be engineered to include two pipes or more depending on offload time.

We also need to plan on it taking a few hours to dock and depart. Again, our rule of thumb for the purposes of our plan is four days in harbor to fill and four days to off load. That is most likely much more than we will need, but a safe and workable rule. Working on ocean conveyance our ship captain/pilot in

Alaska has suggested a total of 7 days per load and maybe 5 days to off load – depending on the facilities in market.

If we find and secure a market partner who is willing to locate a plant at tidewater with direct tanker access, which is our preference, we could reduce the 'in port time' at market to 30 hours plus. This could add another day saved in each round trip. Remember the day-cost of a tanker.

12.3   Shipping – additional notes

As we have pointed out, in the water business "conveyance" is almost always the determining cost variable in product viability both on land or at sea. **It is also the largest uncontrollable cost** mainly because of the instability in the cost of energy. For example, the rule of thumb in the bottled water business is that about 80% of the cost of any container of water is the cost of conveyance and 20% is for marketing (containers are marketing). The actual water in the bottle rarely makes a mark on this cost spectrum. But remember we are a premium 'fine' water, and our brand will change the ceiling in any market – subject to marketing and individual use packaging.

Due to the location of our twenty water sources and the markets (China/India) under development, the only viable conveyance technology proven today is a tanker and as we have pointed out, we prefer a new freshwater V Max VLCC class tanker designed specifically for our use. They travel faster than bags or tugs/barges, with large lift capacities (90 million gallons) and are well established in the international shipping community for long term lease, purchase, operation, and maintenance, docking, etc. Using tankers does pose some marketing challenges, but with the right paint job (white with glacier blue trim) and some creative marketing, we can overcome this obstacle.

The best case is to find a new V Max VLCC tanker that was originally built for hydrocarbons, but never used for that purpose due to a change in specifications or market. We have been advised that such may exist from time to time and are attempting to track them down and determine their long-term lease (minimum 10 years) or purchase costs.

The V Max VLCC can carry 90 million gallons of freshwater or 378,541.178 cubic meters. Modest design changes can convert tankers already built but not yet in 'hydrocarbon' transit, can be made rather quickly to accommodate freshwater.

12.4 Energy costs

As we have pointed out, the cost of energy is the **largest uncontrollable cost variable** in this business. Because of this volatility, we will always try to segregate/identify the cost of energy against all other costs of our operation. We will do the same in the presentation of "day-costs" so that we continue to have a real understanding of costs that offer some **control** based on leases and contracts as opposed to costs that are more variable.

12.5 Using Ballast tanks for freshwater

A concept that has often been discussed but we do not think, based on the new double hull V Max VLCC design, makes sense. We put this in so that you are aware of this idea and our consideration of it because we get questions about it often.

This is an idea developed some years past in the initial consideration of lifting freshwater. A tankers' ballast capacity is about 20% of its total volume of lift. The key is keeping the ballast tanks clean, no oils or other contaminates including seawater. This is generally achieved by flushing and refilling ballast tanks in the open sea but not in a harbor where the seawater is normally not clean. This can be a tricky operation even in calm seas.

Ballast tanks are generally filled with seawater most often in port to maintain balance. Washing out these tanks at point-of-fill with a small amount of freshwater it was thought could make them capable of carrying twenty percent additional freshwater for sale. The problem is they did not adjust for the additional weight of freshwater vs crude. The lift capacity for a V Maxx with oil is 92,000,000 gallons, but for freshwater it's only 90,000,000 gallons due to the additional weight. So, filling ballast tanks is not a viable option, except to return empty and then you use seawater.

12.6 Other Important Cost Variables when Tankers are Purchased

- Life of tanker
    - **30+ years** w/V Max/freshwater design
    - 20 years standard VLCC converted for freshwater (no oil in system ever)
    - 15 years for crude oil
- Volume of water delivered
- Distance of delivery, and return, and time at docks – both end
- Capital cost of tanker
    - Amortization
    - Depreciation
- Capital cost of on-shore or near-shore facilities at source and delivery points
- Permitting and compliance costs at source and delivery points
- Operating cost of tanker (day-cost) Identify the cost of energy for each.
    - Under lift
    - Empty, return
    - In port (load or off load)

Figure 42: Cost Variables when Tankers are Purchased

Of these other variables, other than the volatility of energy, two of the most significant are the length of contract and tanker life. Capital amortization and depreciation of initial capital investment and permitting compliance costs over a ten-year contract versus the twenty or thirty-year life of a new tanker is a significant factor in the cost of water delivered.

For example: If you can pay off a tanker with a 30-year life in ten years, you would have an additional 20 years of operation without this debt, thus increasing profit. Operation and maintenance (not including fuel) costs are generally stable with plannable increases over the life of a tanker. This strategy is extremely attractive for corporate income given the generational nature of this industry.

12.7 Hyundai Heavy Industries – a VMaxx Proposal

We have reopened and continue discussions with Hyundai Heavy Industries to begin serious conversations on the design of a new freshwater tanker, but again they have insisted we present an 'approved design' before they spend any time with us. We understand and continue to research concepts pending our hiring a tanker designer to specifically design a freshwater VMaxx tanker. Our preference is still the Swiss company, Stena Bulk, with offices in Texas as they 'invented' the V Maxx. Hyundai Heavy will then work with the designer and cost out these tankers for build in Pusan. We plan a visit to Texas later this 2023 year for several reasons, but we would like to at least get an agreement towards a water designed V Maxx.

Assuming a 31-day average round trip schedule and 90,000,000 gallons of freshwater using a VMaxx VLCC design per trip to China, the estimated all in **cost** of providing our water to our China market is approximately $0.07 per gallon per round trip, including source water acquisition. Again, this is based on a 31-day round trip, including 4 - 7 days at each end to fill and empty. If that is reduced by better load or offload efficiencies and access, or increased speed, etc. allowing one or two more lifts per year, this cost can be further reduced. We think this is possible but for the purpose of being very conservative in our cost/price estimates this is what we will use. As soon as we are in operation with a solid market partner and hopefully a conveyance partner, we will explore tightening our round trip to add at least another half trip a year per vessel but we hope more.

The overall cost of delivery must also include such variables as shore facilities and onshore systems and permits/licenses, as necessary. With the careful selection of an already developed source, these additional costs are minimized.

12.8   Storage

If possible, we do not recommend this venture get involved in any at market storage in our first five years unless it is **essential** for a Delivery Contract signature - and then only as a Joint Venture with the buyer. In some markets, depending on the margin and dependability/aggressiveness of the buyer, this may make good business sense, but the capital cost and political uncertainty of providing on-land or even in-water storage in China is not a cost that should be applied to this venture at this time if avoidable. Most harbors capable of receiving bulk water (often a reverse flow from a system that provides water to vessels) can provide storage or direct conveyance (pipeline) to the production facility through other systems.

Again, the limitation is most always the size and number of pipes at tidewater. This is a cost factor that initially must be on the back of the buyer/market partner if possible.

Even later, when negotiating delivery contracts with a buyer, we should insist that they provide on-land or in-water storage. (In-water storage refers to the use of Water Transport Bags (WTBs) that can be used for freshwater storage in some markets provided the seawater is cold and stable enough to maintain the freshness of the imported water and local currents will not dislodge the bag.)

12.9 Distribution

The in-market distribution of water to bottled water or beverage markets or industry is the responsibility of the buyer/market partner. Internationally the 5 and 2.5-gallon bottles are the standard for our marketing as well as personalized sizes such as the "pet" bottles in various sizes (.5-liter, 1 liter, and 1.5 liter) and cases are also generally standardized in most markets. All existing bottled water or beverage manufacturers have extensive distribution systems to their markets, or they would not already be in business and considering the purchase of our water.

It is, however, important that we understand each buyer's capacity in receiving and distribution so that we can "push" our bulk deliveries to anticipate their sales based on fluctuating seasonal demand. Water consumption goes up in warm weather and down in cold weather. So, a static delivery schedule may not work for some markets. Again, this is a concern that must be addressed in the delivery contract.

12.10 Development & Engineering Plan

Our site engineering and development costs are presented in the Start Up budget which is the first phase of our five-year budget. Development will not start until we have a delivery contract. If we are providing conveyance, we may need to also design and develop an off-loading system in market depending on the port and buyer we have contracts with. It is likely that the cost of that system will be far less than our systems in Alaska (cheaper labor, materials, and far less compliance) but again, this should be at the least a joint venture with the buyer. We will need to determine that need/cost on a port/market basis once we have a market partner/buyer as it will affect the cost of our water at market.

12.11 Permits/Licenses

The acquisition of our initial twenty sources in Alaska, all though delayed due to COVID and my medical issues now resolved, is already underway. In our design for the acquisition of each of these sources we have worked with our engineers and other natural resource experts in developing solid cost estimates for the full development of our sources. We estimate that full site development will require two spring-summer seasons per site. This is because such work during at least four of our winter months in SE Alaska and not often friendly to such work. If we can work and keep our contractors safe, we will.

12.12 Loading and Off-loading permits

The permits we will need to secure our markets will depend on each port/market. The costs of off-loading and/or docking permits/fees will need to be determined on a port/market specific basis, but we are certain they are within the "other costs" variable tacked on at point-of-sale. As soon as we have a market partner, we will request this information using our market partner and then add it to this plan. NOTE: under table payments are often solicited at ports. This requires a good relationship with the market buyer and his contacts. We do not pay under the table – it's a federal felony even when in a foreign country.

12.13 Shipping Licenses

These are generally held by the tanker owner and considered as part of the "day-cost" of the tanker. We continue to explore other licenses we may need to acquire, but they appear to be port/nation specific and vessel specific. This is one item that we need to address after project financing as it will most likely require on site discussions and negotiations as well as some assistance from a tanker broker and the U.S. State and Commerce Departments who are most helpful with these things. We have included these issues in our market analysis study under discussion with Euromonitor. The first component of their report will deal specifically with regulatory issues/barriers we will need to address for each market when we have a vetted water broker or direct buyer and no less than a draft delivery contract.

12.14 Description of what needs to be done

1. The investors/partners approve this plan or request changes and then approve it
2. Initiate cash flow strategy/schedule for $6 million Start Up funds essential to further develop this plan and put it into action. Payments based on monthly schedules as presented in the approved start up and development schedule.
3. Hire the top crew (4) as full-time employees or independent contractors to initiate venture we prefer independent contractors.
4. Sign contract with Euromonitor to move forward with the presented plan
5. Initiate concept marketing strategies and prepare required materials, media & printing
6. Complete source acquisitions (20)
7. Initiate full global marketing plan/program and start participation in global meetings and presentations **with a display and event staff**
   a. Prepare articles about the company and our successes in delivering high quality *natural* Alaskan glacial water to markets, for submission to trade magazines and papers.
   b. Expand aggressive sales effort to secure Delivery Contracts from markets within all markets Euromonitor finds are interested – ten largest cities in China.
8. Initiate sales effort to acquire as many Delivery Contracts as possible from markets in China, then India. Open some discussions in the Middle East and Mediterranean then Mexico, Southern California and South America - all have expressed honest interest.
9. With Delivery Contracts in hand: for we hope 3 sources, we initiate source development(s) as required to meet demand.
10. Initiate effort to secure conveyance – options:

a. Initiate preliminary design by Stena Bulk for new freshwater V Maxx for immediate production by Hyundai Heavy,
   b. Raise capital to enable a contract with Hyundai Heavy to start final design and down payment,
   c. Initiate contracts with independent conveyance providers,
   d. Select a Tanker Broker to find VLCC tankers available that have not lifted any oil for conversion (accelerates conveyance start by maybe one or two years)
   e. Formalize ongoing relationship with South Korea/Hyundai Heavy/Bulk
   f. If tanker lease is the option chosen, initiate formal negotiations for a minimum 10-year lease of a **new** tanker (clean VLCC class, prefer V Maxx/clean)
11. If necessary, secure additional talent/support to aggressively secure more sales in China, and India. Middle East markets are second priority unless we have a $$ request – if so, we explore sources in the Atlantic (Canada) unless they request Alaskan water.
12. Continue to push tanker purchase/lease agreements for each new delivery contract when signed and vetted (State Department). In the China market one tanker can make 11+ RT delivering w/V Max design 990 million gallons a year. This is the minimum Delivery Contract we will need to cover the cost of this tanker.
13. File notice with the Alaska Department of Natural Resources/Water, Section of expected first export date and schedule. Advise key VIPs and selected media for first export event. This is an international event.
14. Adjust Business plan to reflect what we have learned
    a. Conservation technologies
    b. Development of other water sources in Alaska, Russia, Atlantic, Albania, etc.
    c. Seabed mining of freshwater started in the Mediterranean and other close seas

Figure 43: What Needs to be Done

12.15 It is Time (first three years)

We hope this business plan and their appended materials have given you confidence in our knowledge of all things bulk water around the world. The option to seek other bulk water sources potentially other investors/partners may express interest in, is an open discussion. We would like, unless that interest is from a viable market and willing to be a market partner, to keep this venture an Alaskan global firm. We believe that this new generational business can make a huge difference in the lives of thousands of southeast Alaska men and women all in our economic and political best interest.

One of our commitments is also the development of 'new water' talent in Alaska educating/training Alaska's. These are high paying jobs in water science and its development and conservation. Toward this goal it is our passion to start a college scholarship program with a signed employment contract located in Juneau, AK with AQUEOUS International, Inc. for a minimum of 5 years after graduation/certification. We want to make money, but we also want to make a difference in our world and as many parts of the word who desperately need 'natural' new water.

CRITICAL
We MUST not lose this (2024) field season to ensure we remain in front of all other potential competitors globally. We need to sign contracts with all our scientific experts to ensure we secure ownership of all twenty of our sources within the next 12 months.

The formal commitment of $6M for our first three years, consistent with this plan, is essential. We need to immediately underwrite the work that must begin before our June field work to establish our credibility and allow us to move forward quickly to secure contracts, especially with the tanker industry who will not talk with us until they can see our substance.

Securing Delivery Contracts in at least one market (we recommend China as it is the closest largest and fastest growing) will require travel with onsite face-to-face negotiations, with S. Korean partners and in-market vetted buyers. If not, and a Delivery Contract can be acquired by an investor/partner already in China for this venture, all the better.

To accomplish the first lift of water, once we have a Delivery Contract, we will need to immediately start the preliminary design, costing/financing, permitting, and then construction of the water loading facility for a market we have made a commitment to. Remember, our lead time is best case 23 months due to the time it takes to build one tanker and the likely two seasons of onsite necessary work to get it all done and tested in that time.

It is our recommendation, based on discussions with AIDEA (Alaska Industrial Development and Export Authority), that our infrastructure for loading be financed by AIDEA and built by our contractors. This provides huge political support and market confidence. This facility that we own, can be open to any tanker based on an approved schedule, licenses and permits.

As soon as a Delivery Contract and a source contract are secured, we must start negotiations for at least one tanker (23 months to build with up to one-month sea trial). If the prospective water buyer wants more water faster, we will likely need to lease another tanker(s). This will likely require a deposit or down payment as tanker brokers will NOT get serious unless some amount of 'earnest money' is put towards the deal. It is also generally a good idea to offer two or three brokers the work on potential leasable tankers so that we are not taken advantage of by just one. It also allows us to have a much better sense of comparable costs when negotiating a minimum ten-year tanker lease agreement.

At this same time, we must initiate an interactive process to develop and articulate a strategic corporate development plan for this venture over the immediate 3, 5, and 10 years. This critical management tool is essential in the daily and quarterly management goals and objectives that MUST be agreed upon at the onset of our response to a delivery contract. That does not mean it will not change. These plans are strategies for action with agreed deadlines, not mechanisms or prescriptive devices that constrain management, but we need a clear roadmap of where we are going internationally and the choices that must be made and when at strategic dates. Doing this will require the full participation of our partner(s), project managers, and contracted experts as we determine are needed. We also recommend an independent third party be contracted, who does this professionally, to assist in the development, articulation, and annual review of this critical management tool. When we have a delivery contract that is a 'project' and we need to clearly understood goals and objectives in segments of our growth with each market partner-buyer to move the envelope bigger as fast and their market can accept.

Also, we need to immediately engage our brand or image marketing team along with our business development team. They will participate in the development and articulation of the strategic plan as marketing is an **essential** component of the success of this venture and following phases. The selection and articulation of our image is **culturally a critical decision** in many of the markets we will be working in. Alaska is who we are but having lived and traveled across Asia for many years we must be responsive to branding that works in Asian and other markets.

Where do we operate out of? This is an interesting question about "place". At what point in the development of this venture do we need to establish a place or home office? We do not recommend this question even be asked until it becomes an apparent need to the day-to-day operations of the company. We do not think that it will be within our first couple of years, but it is something we need to discuss. Should we remain in Anchorage or move the main office to Juneau? Until this need is apparent, we can operate as a *virtual company* using email, phones, fax, and Fed Ex and an executive suite (no more than 5 seats) as needed. We should consider establishing a postal address in Juneau while maintaining our Anchorage, Alaska address for credibility purposes. We have seen too many new companies get so invested in a new office facility and then must maintain it, that they commit critical cash flow to a "place" that should be better invested in marketing, sales, conveyance, or new technologies. We will know when this decision needs to be made, but we do not need to rush into it.

NOTE: My preference is an office with an attached work room that allows sufficient space to work with teams and place site maps on the walls along with all essential media to enable global communication regularly.

As soon as a tanker is found, leased and/or ordered, we need to start hiring crew for a cross-sea venture – unless all conveyance will be via a separate company, our preference. Obviously, our crew, if we own the tankers, will not be on payroll/independent contract until appropriate, but we need to start this early because it will take some time as captains and first mates with specific experience and licenses in V Maxx or standard VLCC tanker operation and in our water sources and market ports will be a challenge. With a tanker lease this is not always within your control, but once a tanker is purchased and sometimes even in a lease agreement, you must hire the crew and captain for that vessel all of whom are licensed in that class in the USA and all markets. Our own tanker pilot/captain will oversee these functions.

We have learned that when dealing with other nations and/or cultures it just works better **if you have a mixed vessel crew** representing both ends of the trip. So, if the first Delivery Contract is in China, at least a third of the crew should be Chinese. There are many reasons for this, but when the vessel reaches market (in this case China) having crew members from that country enables things to work far more smoothly and often at less expense – and in the local language.

An ongoing priority task will be market development. She is already working on her strategic plan per market and water source along with all global water events that require our presence (note below). These are keys to our success.

Again, this is a part of the strategic planning process, but it also **requires at least a one-year advance commitment** to several "sales" opportunities such as the World Trade Center, the World Water Congress, International Desalination Association (IDA) Trade Show(s), International Bottled Water Association (IBWA) Trade Show, and all international or trade publications that touch water in any way. Our preferred target are magazines and events that are focused on Asian markets, and not just partners be always consumers, and there will be many.

Securing new Delivery Contracts, although in some ways regionally strategic, will be an ongoing task, and after source acquisition, our highest priority. This is directly tied into our market research, marketing, and business development, but once our image and brand media are out, it is highly likely we will be approached by other prospective markets that are looking for high quality water for bottled sales, beverage stocks, or other types of freshwater needs – one or two are most likely from B.C. Canada.

Based on the success of the above listed actions, and well before we run out of water from our initial sources (which will never happen), we need to initiate the acquisition and development of other water sources in and outside of Alaska. At that point, which we think is likely year 5, we suggest we begin immediate discussions with sources we know in southeastern Russia, western Albania, Turkey, etc. Such negotiations will take many months or longer and the type of contracts, licenses, and agreements that will need to be drawn up and put through each sources system will require a focused and multifaceted effort and legal team.

If we can secure a water export license from Albania with the help of a market **Source Partner** (a company in a country who takes responsibility for all things in their country), we will then have several extraordinary sources to serve the Middle East and all Mediterranean markets including Corfu, Italy and North Africa.

At about this same time, roughly five years, the next phase umbrella/**global** business plan will need to be completed, building on our success and we hope few if any failures, our markets and future opportunities will map out the next 2.5 decades.

In the water business, people think in 50+ year strategies. Most sources and markets, at the municipal water supply level take a **minimum** of 5 years to develop. The infrastructures and conveyances are thought of in 20 to 30-year blocks for tankers, and 50-year blocks for other types of infrastructure. Therefore, many Delivery Contracts now run 50 years (some 99+ years) in length. This is also the case due to the constant concern about security of source for every dependent economy. The security issues we will only discuss in total confidence based on my experience in the Middle East, north Africa, etc.

It is important not to underestimate the **"narcotic"** impact of water. Once an economy becomes "addicted" to a water source and, as a result it prospers, it becomes more demanding. It is fascinating to watch an economy build such addictions over time. It is generally not something "they" realize at the start, but they become aware of it anytime there is an interruption of supply or a sharp increase in cost and then panic- like a drug withdrawal.

12.16   Items to be done to achieve goals

Listed are the specific actions over the course of this five (5) year business plan that MUST be accomplished to achieve our goals.

COST
Cost of corporate development
- See attached recommended budgets for 2021 - 2025

Cost of securing/ownership of first 20 AK water source(s)
- See included start up budget

- Initial consulting/employment agreements, etc. MUST be signed now to secure these professionals **before** they get picked up by others (a common problem in AK). ($)
- We recommend an initial employee payment of $5,000 so we can sign essential contracts for talent, travel, compliance, etc.

Cost of securing the first water sources in SE Alaska
- See attached budget

Cost of securing <u>first</u> Delivery Contract
- **China** (US State Dept/CIA advance approval)
    - Estimate minimum of $100,000 including cost of entry
        - Strategic market analysis ($50,000)
        - Travel/accommodations ($)
        - Gold Key Service/U.S. State/Commerce Depts. & CIA review
    - When we secure **a market partner**, this cost may drop precipitously

- **India** (after we are 'established' in China via S. Korea)
    - Estimate minimum of $100,000 including cost of entry
        - Strategic market analysis ($50,000)
        - Travel/accommodations
        - Gold Key Service/U.S. State/Commerce Depts & CIA review
    - When we secure a market partner in India, this cost should be less.

- **Middle East**
    - Unknown now.
        - Strategic market analysis ($50,000)
        - Travel/accommodations
        - Gold Key Service/U.S. State/Commerce Depts & CIA review
    - If a market partner from this region joins the company, it will make a difference in this cost.

Figure 44:   Costs to be Identified

12.17 How proven is the technology?

The use of large tanker vessels across the open seas has been a proven technology for generations in almost every global market. The oil tanker is the best known, but other tankers that carry chemicals or food liquids (olive oil, fruit juices) in bulk have been in operation for decades.

The use of large tankers and sea barges for transporting bulk water is no different than transporting oil or other liquids other than the cleanliness that is essential for the water. Converting old oil or chemical tankers to water has been studied and found not economical due to the limited remaining age of a modified tanker.

12.18 Options to each significant task

The only reasonable alternative option to the projects and their components outlined in this plan would be to **not** start with the water sources in southeast Alaska. This would then insert a significant level of risk because, as of today, Alaska and America are the only places on earth where we can 'own' the water we get appropriated to us. It becomes property once permitted. Additionally, almost any other foreign water source will have a higher degree of legal and political uncertainty over time (30-year minimum). We also know from other experiences, such as those with World Water, SA, that the development and licensing of other bulk water export sources can take considerable time (3 to 5 years) and costs that may then change again in just months. Outside of the United States such sources do not have the same level of legal rights or contract law protections and are most often subject to political changes outside of the control of the company licensed to export the water, or illegal requests. The only option is a market partner who is well established politically to protect 'their' investments and ours.

We believe it makes a great deal of business sense to start with the extraordinary high-end pristine *natural* Alaskan glacial water from Alaska as are being secured in SE Alaska sources for retail beverages and industry and then explore other potential water sources in Albania, Russia, and the Atlantic. Once secured they are contractually far more secure then in almost any other region of the world including Canada. Just a decade ago, Canada/British Columbia arbitrarily canceled a dozen export permits when the Provincial government changed hands resulting is deep losses to several American companies.

13 MARKETING

13.1 International Meetings

If this company is to be seriously considered "in the market" it must participate in no less than three biannual meetings and at those meetings we must be a fixture on the "showroom floor" and its best location will be a corner. This is expensive, but this is where we get introduced to the world and where we will often sign contracts.

1. The World Water Congress

This is an international body of national governments, NGO's (non-governmental organizations), of which we are careful, global water businesses and organizations concerned with the pending global freshwater crisis. It meets every other year. Most of those who participate at the diplomatic level are having trouble just understanding their immediate and long-term freshwater challenges, and do not understand <u>solutions</u> to their crisis other than "more government" or more UN money. But they all come to the trade shows and talk with companies about their needs and thus offer an extraordinary opportunity to make the pitch and initiate a sale or Delivery Contract. To be creditable we will need a presentation that shows who we are and what we are or will be doing. We will also need to be at some level of sponsorship so that we can get information into the Agenda Book to entice participants to visit with us on the trade show floor. Remember, these contracts are multi-year agreements, some 99 years.

Additionally, **all** the freshwater "providers" globally are on that showroom floor. If you want to be taken seriously as a global company and you are not in this trade show, you have a serious credibility problem that you may not overcome. It also offers us the opportunity to take a serious look at any developing competition and connect with possible partnerships.

2. The International Desalination Association - World Congress and Trade Show

This is the other global event we must attend. It is hosted by the IDA&WR Association and is far beyond just a gathering of industry folks involved in desalination or water treatment. This is one of the major opportunities for companies to showcase their technologies and generate new contracts.

Our participation in this event will cause significant attention toward us. This is not a cheap event and participating on the showroom floor takes at least a year's preparation including reserving the showroom floor, hopefully corner space. But the potential returns on this marketing investment can be significant. Again, we will need to participate at some level of sponsorship to get our information into the agenda and preconference materials so that potential clients will know we are there, we are a player, and where on the trade show floor they can find us.

IMPORTANT: We are the competition to desal, so we will be welcomed with raised eyebrows. But this gives us an opportunity to sell 'mix' that will help the desal industry raise the value of its product at less cost and certainly less environmental and early human impact. The theme of our booth might be something like, "Just add Alaskan glacial water" in a visual that is immediately eye catching. We could also take a part of a glacier and collect bets on when it collapses and melts. I did this once in Florida at the annual state fair, and it was extremely popular and got excellent media.

3. International Bottled Water Association

This organization meets biannually as well. Since we are starting with the sale of a high quality unique *natural* glacial water from Alaska, we MUST be at this event and have a location, we would hope a corner, on the showroom floor. This will not only afford us the opportunity to sell our water, but also to interact with hundreds of potential buyers across the globe. We will also meet many other water technology companies who may have an application to our efforts that can benefit what we are doing. IBWA is in our opinion the best international gathering of "players" in this global water market.

In addition to these three global opportunities there are almost monthly, regional events that we should participate in. As we move into markets in these regions, we can coordinate our sales trips to take advantage of these marketing opportunities. Obviously, our first regional show must be in Asia. Then the Middle East, Africa, California, and Mediterranean states will be our next targets. We will also want to do an event in London because it is generally the banking center for almost all Middle Eastern businesses.

> SIWW, Singapore, June 29, 2009 - (ACN Newswire) - The second Singapore International Water Week ended on a high note for many local and international water companies who sealed deals totaling **$2.2 billion USD**, nearly six times the amount achieved the previous year. More than 10,000 attendees from over 85 countries/regions visited the week-long event.

The next International Water Week will be October 3 to 7 in Yogyakarta, Indonesia. I would expect 10,000 people there and every water company in Asia will be on the show room floor. Given our new and revolutionary presence, we will request an opportunity to speak at the event. Yes, we will have to pay for the time, but with our marketing we should have a full house.

13.2   Publications

Global Water Intelligence

This is THE global point of information we must be in. This is a global inventory of all things water that is updated and sold (about $5,000) every year. If you are not in this series of reports, you are NOT in the business. This is also an annual expense for the company as it lists all projects, vendors, developers, water owners, etc. across the planet. It is a MUST have in doing any water business because it has all the information you need to contact buyers and they have all our information too.

Once we get our image nailed down and the website is updated, we need to send our "marketing package" to Global Water Intel for inclusion. They will then review it, often call with questions, and then vet the company before putting it in their various reports. If we want to be highlighted in this publication an advertisement is essential.

We will be included in all the other water business publications once we are members of their organizations. These include the International Desalination and Water Recycling Association, the International Bottled Water Association, American Water Association, U.S. Water News, etc. We will need to put together a series of articles for these trade journals and magazines that will tell our story and keep prospective buyers updated, thirsty, and wanting to visit SE Alaska.

The World's Water

is published biannually by The Pacific Institute for Studies in Development, Environment, and Security. The key person in the development and writing of this book is Peter H. Gleick. Peter, the founder and President of the Pacific Institute, is an anti-growth advocate, but produces a book that is now a must read for anyone in the global water business. It has one of the most comprehensive inventories of potential water markets published. If Peter likes you, and we have been communicating for some years, and thinks you have something to offer in resolving the global freshwater crisis, you might do well in his treatise. This is an important book and group of writers we must take seriously. If they do not like you and they decide to go after you, you need to be prepared and willing to spend what is necessary to overcome their misjudgments. This is one of the reasons our marketing company must be one that understands water and all international issues around water.

Fine Waters

The connoisseur's guide to the world's most distinctive bottled waters. Alaska has a fine water presented on their website - The Alaskan glacial water originally produced by Rob Gillam (long time friend and potential investor), imports to markets in China, India, and the Middle East and beyond and opens the door for our "feed stock" to be recognized in this unique publication. Working this book, used by all the high-end bottled water marketers, into our global water marketing strategy ensures recognition of the unique value of this product and should open new sales opportunities globally in all high-end markets. This book also ranks and analyzes high-end bottled waters of all stripes. Recently they noted that Parle's Bisleri who once monopolized the market is now vying with Nestle, Coca Cola, PepsiCo, Manikchand, UB and Britannia. Getting into this publication will require the skills of a very competent marketing team which we have, and a product that is worthy and soon available around the Pacific.

13.3 Presentation

The development of our international image is very, no let us be honest, **critically important**. How we do this and who we do this with, especially in some markets/cultures, will determine our access and our success in many markets. In this area we have previously worked with a company that: 1) understands water and global markets, 2) has an international reputation for culturally sensitive images and presentations and 3) is multi-talented in how they present who we are, what we do, and why anyone should care.

"How they felt"

In addition, we have ready to join the team someone who we have worked with in other ventures and is an exceptional creative and marketing talent. She brings emotion to marketing, something many do not. As she has said many times, *"They remember how they feel more then the words they read.'* She works all the time and has called the Kenai Peninsula home since 2015. It is our intent to put her in the Vice President of Marketing once we see how she does full time with us. She will most likely do an extraordinary job as a member of our team. She will coordinate with other branding entities, such as Pixel Farms pixelfarm.com, to propel AQUEOUS International, Inc. a head of the pack.

"Since 2015, Asian water markets have grown over 15% per year in 21 Asian countries with China leading the way. This trend is steadily increasing. The need and desire for clean, fresh water for industry and human consumption is always great, and now water from pure sources in the northern US has become an active trend and focus of these large-population markets, for health, purity and even status", she says. She brings years of experience promoting businesses in the unique, discreet Asian marketplace, and can position AQUEOUS International, Inc. for success as the world leader in the pure-water market.

The central message, especially with the water products from Alaska, is critical to establishing "top of mind" for both businesses and retail buyers. One of the more interesting marketing concepts developed in the early days of the limited bottled water export business in Alaska was how to compete with Avion, known as the top-quality bottled water product, globally. Much of their reputation is built on name, but also the notion that their water is better than anyone else's. Why is this? Is their water source really that unique and pristine? The answer is no, **it is just marketing**. It has always been the Avion bottled water strategy to be at the top of the price for bottled water regardless of cost or market. Buyers are willing to pay more for what they believe to be better – even if it is not.

To compete with the Avion illusion of quality, one bottled water manufacturer in Alaska developed the following story.

"Would you rather drink a bottled of water taken from 32,000-year-old pristine glacial ice in Alaska or water from an aquifer that has been pissed in for over 2,000 years?"

The Food and Drug Administration (FDA) did not like this "marketing strategy" but it really does tell the story.

What we will immediately need to develop is:
Our brand/image which will include:

- Selection of a name (should it be AQUEOUS or something else?)
- Logo
- Website expansion (must be interactive i.e. Pixel Farms forte)
- Business Cards (with various language translations on the back)
- Letterhead and all mailing pieces (presentation folders, stationary & envelopes, labels (essential for international mailings), etc.)
- Our Story – shorten what we have in this book
- Video presentation(s) in various languages – show sources (3 minutes max)
- Illustrated business and strategic plans that can be provided to key persons

Figure 45: Immediate Development Needs

14   BUSINESS DEVELOPMENT

To maximize potential sales, the approach will be targeted to 3 primary areas within each country/market:

1. Large and mid-capitalization beverage manufacturing companies focused on alcoholic and non-alcoholic beginning with those who **need large** quantities of fresh and pure water. Our water solution will enhance the quality of their respective products which they can use as a new "Edge". For example, " Brewed with the purest and cleanest water on earth". A strong selling point as water becomes scarcer and contamination grows globally.

2. Large and mid-capitalization cosmetic companies. Our product should enhance the quality by decreasing harmful minerals and chemicals used in the desalination process which may cause allergies and skin rash and irritation.

3. Various governments have expressed the need for large bulk freshwater for their respective populations.

   *critical we do not disclose too much before we have an NDA signed.

The assumptions made here reflect the ability of large and mid-size companies to make relatively quick investment decisions as compared with those of government.

Tailored Marketing
Each marketing strategy will be tailored by simply altering the "Digital Pitchbook" to reflect each company's respective "Core Strategies" and corporate focus and theme.

The "Digital Pitchbook" will be sent electronically after the first contact is made with senior management and then followed up with a face to face meeting to close the account.

The pitch will be designed with risk mitigation strategies to include "Options" which will allow a greater understanding for each client of their fixed costs which they will be doing business under.

15 MANAGEMENT

15.1 Management Team

Ric Davidge, MPA/PM Founder/Chairman is coming in as the 'talent' investor and will function initially as the leading visionary member of the management team, project development team, and a corporate director. He has suggested three other positions he believes necessary with talent that can jump start this venture based on their experience, professional integrity, creativity, knowledge of water and its conveyance, new water technologies, sales, industry contacts, etc. Davidge will involve himself in all and any aspect of the company with a key focus on securing new water sites and international marketing w/targets.

Chief Operating Officer is responsible for all operations including site caretakers. Will oversee all personnel including independent contractors; source site development and related issues, infrastructure relating vessels and sources; vessel acquisition, testing and operations; Financial accounting and business administration, reporting and forecasting.

Marketing Director will dive immediately into all things marketing. Rebuilding our website, structuring our global brand and a list of labeling options for buyers, etc. One of the most important creative things she does instinctively is putting emotion into image and narrative.

The key immediate needs are water source acquisition (in process) that requires significant site design required by the State of Alaska to secure our 'ownership', loading design, along with sales and marketing, and business development. Each of these functions demand either a full-time team member or consulting contract, phased in as we move forward.

Fabric Engineer will be placed on contract **if** we find a need for Water Transport Bags as in sea storage in some markets. Sri was the designer and 'inventor' of the fabric we had developed for the Mediterranean markets.

15.2 Project Development Team

Under contract structured relationships with people or companies, mainly in Alaska, to meet the immediate needs of the company and then assist in our accelerated growth. This is now in process. PND Engineering is the project engineering team selected due to their many decades of work and experience in Alaskan waters, both fresh and salt. Mr. Davidge has worked with PND in many previous varied ventures/functions and has found them 'the best' most experienced in sea and rivers in Alaska.

Immediate discussions with a tanker company for the initial design of a V Maxx specifically built for freshwater and our operations. This is essential at start up due to the time it takes to realize a licensed seaworthy freshwater tanker. This will be led by our tanker pilot/captain, (Richard Gurry) who is still active in the tanker business, but now available to the company.

15.3 Contractors

Stena Bulk

is the designer/builder of the original V Maxx VLCC tanker we have worked with. We will need at least DRAFT design and construction license agreements if we want to use this design in the S. Korean shipyards. We believe that the relationship between Stena Bulk, Hyundai Heavy, and AQUEOUS International, Inc will develop a close team for tanker design, development, and correction as needed.

Hyundai Heavy Industries

is one of the companies that we would like to build our first tanker. We have established discussions with them. They want us to provide a designer/design for the first freshwater tanker. We want to ensure total control over any access to these designs during and after construction. All designs will be registered and protected by federal law. Hyundai Heavy Industries are fast, competent, open to change and innovation, often make knowledgeable suggestions, etc. Mr. Davidge also speaks some Korean having lived there for over 2 years. We will of course have our own interpreters approved by the World Trade System and our state and/or federal or Commerce Departments. We have used this process around the world, and it is always well worth the investment.

Ultraviolet Light (UV)

This is for water treatment systems for the tankers allowing us to deliver "drinkable water" at market. The cost effectiveness of this is yet uncertain, but we are continuing to pursue this possibility as it will greatly increase the value of our water for wholesale buyers. There are no other solutions we have found that meet our objectives of no chemicals but drinkable at market without any filtration.

Other contractors are being gathered based on the need to design and then construct the loading/off-loading facilities in SE Alaska, and if necessary, in markets.

We believe it essential that we have market partners/investors. These are substantial investors who come from each market we are interested in serving. Such a relationship gives a much higher degree of integrity to this venture in-market and keeps us out of local/national political problems in market we do not want to deal with – especially the request for under-the-table payments.

Planners and Doers
The Board of Directors will be established by the investors, but we need a working group of "players" who can make decisions on a regular (likely weekly) basis, especially over the first eight months. Given the speed of this opportunity, at least in securing the water sources we have inventoried and are in the process of acquiring in SE Alaska, and the delivery contracts in China, South Korea, and maybe India and California/Nevada, we have just months to get a lot of work accomplished once the venture is financed per agreement. Keep in mind that our filed work is limited due to the weather, but we have a particularly good pilot in Juneau we trust to get us in and out safely.

Weekly progress reports will be provided to every investor or as they require. In past work we have found quarterly reports more substantive because you have accomplished something measurable rather than just moving in that direction. It's important to remember that our SE Alaska work season is only about 5 or 6 months long. Within that window we need to get everything we can done 'on the ground' so we just 'follow the paper' over winter. We also want to initiate our BRAND marketing just as soon as we have a Delivery Contract as then we know the market we are targeting, and we want to start our media 'drip campaign' no less than 8 months before anything is on the shelf or in market.

16 FINANCIAL PLAN

16.1 Legal Structure

Within AQUEOUS International, Inc. we recommend the establishment of an Alaskan company that can function as a (in 5 years) global holding company (AQUEOUS Global Holdings, Inc. is our suggestion) legally located **outside of the United States**. We suggest Luxemburg based on previous experience and the preference of former international water partners, especially in the Middle East. But we must maintain the Alaskan image for all our marketing even when we explore and secure foreign freshwater sources with an in-country partner.

We also recommend, once the global holding company is established and we are seriously serving a market-based project, that each project or groupings of them, be structured as separate legal entities or profit centers with majority share ownership retained by AQUEOUS International, Inc. This allows the greatest management flexibility in the development of each source and project while at the same time limiting the exposure of the parent company. Appropriate guidelines and approvals will be necessary for each "startup" at the Director's level, but once those decisions are made, especially if dealing with a foreign market or source partner, our experience is that this is the best legal and operational structure with the least immediate and long-term exposure.

All of this is of course subject to negotiation and discussion, but we wanted you to have some idea based on our global experience in water.

16.2 Capital Structure

AQUEOUS International expects, at start, to retain no less than 55% of this venture based on its accumulated knowledge and expertise in this matter and expected contribution to enroll in overall operations. As shares are assigned for various reasons as we grow, AQUEOUS International will increase the total number of shares and contribute its appropriate share in moving the venture forward.

We like share options for senior management as it is the best performance motivational tool in hiring quality talent, keeping them, and getting them to have some skin in the game. It just works.

We may find market partners who want to invest directly in the parent company. If their market is exceptionally large and they have the capacity not only to sell in their market but also to expand into others, they may be a welcome market partner. The only concern is their function in market with their political associates (possibly hidden government partners) as we do not want to have any exposure to such a vulnerability. Been there done that. We will not be owned by any government. Lessons learned.

16.3   Milestones and Objectives for multiple funding rounds

This Phase A plan business plan is only for the first five years of this venture. Yes, we project out 50 years, but the overall purpose ***now*** is the establishment of this venture in the global water world of business through the immediate acquisition, export, and bulk conveyance of very high-end *natural* Alaskan glacial water. The first financing will require no less than $6M US for this "start up" to acquire our sources, design the 'harvest' technology to the vessels, initiate our conveyance options, and secure our markets for a minimum 30 years. This does not assume any major capital outlay for tankers or exporting/ importing facilities in this Phase – although that is possible depending on the Delivery Contracts we secure. We believe it makes better sense to capitalize each of these "investments" as the company moves forward with apparent success.

It is also within the scope of work of the first five years of this company that an updated business plan will be developed with real tested numbers and 'in water' experience to explore a range of options for growth including other water sources closer to Atlantic markets. This may include new technologies such as those that will be necessary for subsea freshwater vent exploration and development. As the Chairman requires, the team will review the business plan and adjust it.

16.4 Budget

Initial budget is built on the assumption of $6M for the first three years as illustrated in the plan.

Employment Contracts

Ric Davidge -Director/Management	Full time immediately
Chief Operating Officer/Ed Hahn	Independent Contract
Director of Marketing/Laruen Di Scipio	Independent Contract
Richard Gurry, tanker captain/pilot	Independent Contract
Business Development consultant/searching	Contract
Sri Tupil, WTB contract as needed	pending demand

Figure 46: Budget- Employment Contractors

Sales Travel: We recommend we start in South Korea towards China due to relationships already contained in this venture. Once we have the Euromonitor reports on markets and vetted buyers in South Korea and China, we must do face-to-face work especially in cultures where 'relationships' are critical in developing any serious market partners. Although we will start sales/marketing at three well known large coastal cities, we will quickly move into new markets in northern China as well as others. We have at least ten huge markets in China to get us moving – the quickest.

16.5   Initial Investors

AQUEOUS International, Inc. an Alaskan Corporation is an investor in this venture bringing enormous knowledge, experience, and potential Intellectual Property assets to the venture. Ric Davidge, Chairman is the largest shareholder (100,000 shares) of this corporation and currently the only investor. There are no other shareholders although some have asked. I'm picky!!!

16.6   Capital Investments

Once we secure our first bulk export license and delivery contract there is one design and 3 immediate capital items we will need to initiate – <u>subject to possession of a Delivery Contract.</u>

o	SE AK water harvest/loading facility	estimated $10-12M (requires two seasons)
o	VMaxx design for freshwater	estimated $250,000
o	V Max VLCC class tanker S Korea	estimated $150M (23 months)
o	Market offloading facility (optional)	estimated $1M (18 months) per site requirements
o	We prefer this be the market partner's responsibility but need this in budget just in case.	

Figure 47:   Capital Investments

16.7   Intellectual Property Investments

Water Transport Bags (WTB):   Bags are still in use in some areas of the world in small scale, but the advancement of bag fabrics and towing technologies continue to advance often separately. An American company has developed by far the best poly fiber fabric for WTBs that can last ten years. Then this fabric can be recycled for roofing material. In almost all cases the life of older types of fabrics has been seven or less years. This makes a huge difference in the economics of bags. Two of the supporters of AQUEOUS International were key players in the development of this technology. AQUEOUS International is often called (most recent is Chile) about the application of WTBs in a wide variety of uses. There is significant opportunity here to capture and "own" some of these new technologies that can generate significant return on investment over the years.

Subsea or Ocean Floor Freshwater Mining: For most people, including many water professionals, this is an unknown opportunity. In some parts of the globe cold or warm freshwater comes up from the bottom of the sea. The quality of this water, due to its surface to subsurface percolation, often through limestone, is extremely high. There are specific opportunities in some sections of the Mediterranean for example and small adjacent seas to capture this water and deliver it to local markets. *In most location this water is NOT owned by any government.* The technologies in the mining and capture of this extraordinary freshwater are available for the right company. These are essentially "free waters" as they are not <u>owned</u> by any nation – at least not yet. Claims can be made through international governmental bodies, but no one has really had the capability or vision or understanding of these freshwater sources as we do to get in this game. This is a serious opportunity to capture and own a unique quality freshwater product for conveyance with extremely limited cost especially to markets in the Middle East, north Africa, and all Mediterranean markets. We already know where some of these vents are and how to find others.

IP Acquisition Strategy: The development of new technologies in the global water business is rapid, dynamic, duplicative, constant, and global. We recommend a special team, within the new Holding Company, monitor all globally filed patents using existing search engine software to highlight new patents that may be of interest both in specific technologies and in new processes. AQUEOUS International Inc. did this for several years, and in that process found many opportunities with struggling inventors, including international universities, to acquire new or emerging technologies that have immediate and long-term value in the water world. This is a very smart investment during this phase in the development of this company. It also ensures that we remain in front of our competition with new technologies. Again, this is a very dynamic field. Once it is known that we are looking at ALL water related new technologies, the investor(s) will approach us before even publishing their patent. It happened once, it can happen again.

We believe that with the use of freshwater VMaxx VLCC tankers, we will significantly lower per unit costs. Additionally, designing in the UV water treatment system will increase the value per unit at market, and we may be able to realize 11+ round trips (China) per year with new vessel technologies we have explored and are now being used, that increase vessel speed.

We have initiated a targeted water wholesale cost/price elasticity study for China with Euromonitor which is included in our startup budget. We have some initial cost/price numbers from a wine importer (now almost 20 yrs.) into China that are very encouraging. Once the China report is completed, we should move into a similar study in India. Remember, we are only comparing them with mid to high-end wholesale water and retail product lines. We do not yet have any prices for manufacturing such as cosmetics and computer components who have expressed interest. These will be further researched once we are underway.

16.8 Detailed cost and profit breakdown/ Project Proformas

As we move forward, we may add some additional numbers we will need to collect and integrate into this strategy before we are fully comfortable with our attached breakdown for the company. To accommodate this, we have allowed some margin in our numbers. The closest we are with the standard VLCC tanker cost and profit breakdown solely based on crude oil lift, that is included in this plan. Although we have engaged with Stena Bulk, we do not have the VMaxx VLCC 'established' freshwater cost numbers yet. We cannot obtain these numbers until we have formally engaged a tanker broker and designer. We believe that the Stena Bulk office in Texas is our best choice. We have been assured that all the costs for a VMaxx are lower than the standard VLCC and its performance far more attractive. That is the reason it was designed and built.

17 LEGAL

17.1 Water Rights and Contracts

All water rights necessary for the 20 water sources in southeast Alaska will be acquired by AQUEOUS International, Inc. within the next twelve months from the Alaska Department of Natural Resources once we are funded. Some are already in process and protected 'under paper'.

17.2 Ability to Export

The water rights and formal determinations by the State of Alaska that these waters, as defined by Alaska law, are "excess to the needs of the hydrologic unit" which essentially includes the entire SE Alaska hydro construct, from which they are taken. That means we will never exceed a water sources seasonal discharge. Therefore, we have been careful in our seasonal take to try and stay within the annual seasonal margin of each source. This is generally 10% to 15% of the annual discharge. Remember, this hydrologic unit (SE AK) is about the size of California and is enormously interconnected with all components of the hydrologic cycle. Then when you add glaciers and ice fields and seasonal melt, you have one of the largest and most complicated hydrologic systems on earth. It is unlikely that these determinations will be challenged as they are very conservative due to the structure of Alaska state law and water appropriation regulations. There are no federal permits involved now in these state determinations as the state owns all the water, ice, and snow in Alaska. No federal Environmental Impact Statements (EIS) or Environmental Assessments (EA) are required. We are now collecting all federal permits, EIS/EAs for all FERC approved hydro systems in southeast Alaska of which there are at least 30. These are then added to each of our applications for water rights – at extraordinarily little cost to our effort. 'Referencing' a federally approved EA or EIS to a facility in our source is a well-accepted practice in Alaska and federal permitting. Only the request of a federal permit or some specific federal action, and we have stayed away from federal lands, that could trigger an EIS requirement.

This does not mean we do not assess the environmental impact of a 'take' per season. We are ahead of the state permitting with these strategies and it is welcomed by the state.

17.3 Environmental Issues

17.3.1 Local

The level of attention to the environmental issues in past water rights appropriations and determinations is well established and generally accepted in Southeastern Alaska. This region of the state, formally designated by the USGS (United States Geologic Survey) as a <u>single hydrologic unit</u> and recognized by the state, is a hydrologic system about the size of California receives more precipitation (rain/snow) then most areas of the planet. This does not include glacial melt. Precipitation here is not measured in inches, but in feet. In fact, an island in this region is either the single <u>wettest spot on the planet</u> or the second wettest annually, competing with sites in Hawaii.

It is always possible for someone to file a lawsuit and claim that the environmental implications of these bulk exports have not been fully considered or understood by the State of Alaska. Given the work of the State of Alaska in preparation of these water rights and Export Permits, and the history of such claims, it is not likely such could prevail. But, considering this possibility, we work extremely hard to keep a low profile. The key is that this entire region is one hydrologic unit as it should be and that the enormity of freshwater precipitation, (rain, snow, ice) and glacier melt is honestly unimaginable to almost anyone. The taking of 100 billion gallons a year from this area in not in any way significant with multiple dozens of lakes and glaciers discharging trillions of gallons a year into the sea. Our take is not even a drop in this ocean.

17.3.2 International

In considering the environmental implications of bulk water exports, it is important to understand the recent effort by some radical Left/Green groups to claim water as a 'human right'. And then, some claim that such conveyances should not be allowed as they "harm" the seller's location, which is in direct conflict with their concept of water being a human right. This was tried at the UN and it failed quickly. But we keep our ears to the noise, and it is one of the reasons we quietly participate in all international meetings that address all types of water issues.

Water has been conveyed hundreds and thousands of miles for generations. Those who wish to litigate to suggest some environmental harm on the side of "human rights" then deny those humans in any market the 'human right' to survive in the face of serious clean water shortages.

In some parts of the world this may be an effective strategy, but in Southeastern Alaska it is not. The simple magnitude of the annual precipitation through snow and rain fall; glacier melt, snow, and other ice melt is beyond the ability of most to seriously understand. This is especially true for people from areas of the planet who think 7 to 10 inches a year is a good annual precipitation level. In many areas of SE Alaska, the annual precipitation is well over 200 inches – every year.

The tradeoffs of desalination and bulk water conveyances are not even close to being equal either in energy consumption, "carbon footprint" or human health value when consumed. Bulk water conveyance, even in large tankers, consumes far less energy than existing desalination plants (RO) and eliminates the need for "onshore" facilities that often consume significant and valuable fish and wildlife dependent wetlands or high value ocean frontage.

Alaska water law (Title AS46.15) is considered one of the best and most contemporary in the western United States, again most of it authored by Ric Davidge when State Director. In considering any appropriation of water for the purpose of export, the law requires that sufficient water be 'reserved' for fish and the future needs of the state <u>before</u> water can be removed from a hydrologic unit. The "public interest" must be considered including the effect of an appropriation on fish and game resources, public health, water quality, public recreation, the economic activity resulting from the appropriation, and the effect of a loss of the water for future needs. Additionally, the rights of prior appropriators, with adequate means of diversion, the use must be beneficial and give benefit to the applicant, any harm to others must be assessed and consider the intent and ability of the applicant to complete the project. This must be determined and considered, and any effect on access to navigable or public waters must be reviewed.

In addition to the considerations listed above, any person, including state agencies such as the Alaska Department of Fish and Game (ADF&G), can apply for an instream reservation of water for fish, wildlife, public recreation, water quality, sanitation, navigation, and transportation purposes ensuring additional opportunities for resource protection. This is in fact what happens in the state's formal determination that the water approved for export is "excess to the needs of the hydrologic unit". But don't forget that the sources we have assessed have some range of grade or pitch that doesn't allow any anadromous fish to reach into the source above high tidal reach. Our take points are well above natural discharge for any source.

But what are the open questions, yet unanswered? What is the environmental impact of an appropriation of water, in bulk quantities, at the point of discharge on saltwater estuaries? This is a reasonable question. In some cases, there is no real estuary due to the nature, grade, or pitch of our sites. In other sites, the extraction of freshwater, which is now being discharged into the ocean, would remove an **<u>unnatural</u> or man-made discharge,** and thus return salinity to a pre-project level. This is the case for several sites in southeast Alaska that have hydroelectric facilities. In most cases this is a site-specific impact question that we always address in our applications. Appropriate environmental studies or determinations are completed as a condition of water permits for the purpose of removing water from a hydrologic unit i.e., all of southeast Alaska.

Additionally, all sites identified to date as potential export sources fall within the Southeastern Alaska Coastal Zone. But the former strict review requirements of the Alaska Coastal Management Act are no longer in operation by action of the state legislature as they were too restrictive of any responsible development. But we use this old requirement on our site and water rights application packages as this provides the government reviewer with a good reference as they ask questions. We answer all questions BEFORE we apply and they ask. This has always been a phenomenally successful strategy in our applications.

The Alaska Department of Natural Resources (DNR) has already facilitated meetings of all state and federal agencies on several bulk export applications. Only one application has been approved, to the City/Borough of Sitka. Sitka, as we have pointed out, does not have the ability to load **any** tankers large enough to make the export of their water financially viable at almost any price in any market. The only viable technology for the Sitka source is WTBs but the cost of loading them, towing them out to deep water, and pumping the water from the WTBs to the tanker is beyond doubling the cost and thus its financial capacity.

It is important to understand some of the numbers. Removing 225 acre-feet a day (one standard VLCC tanker) from a source, logistically is exceedingly difficult, and would require the appropriation of 82,125 af/y from that source. If the annual discharge rate of the source were 500,000 af/y, the removal of this amount of water would be less than 20% of the total annual discharge. Because seasonal flows fluctuate, the State of Alaska does not appropriate only based on annual flow, but rather seasonal flows. Therefore, the State of Alaska believes the industry will need to use several sources, reducing environmental impact on any single source especially in the event of regional draught – which is rare. All applicants for bulk water exports from the State to date have made it clear they are not interested in any source that will conflict with existing uses such as commercial fishing (salmon) or public recreation.

On the other end of this water delivery system are other important environmental questions as well. As most experts agree the serious problem in the southwestern states of the United States for example threatens its population and definitely its business growth. Often during interviews, we are asked about "our" collective responsibility to these pressing issues. We must in all honesty answer that this is not an issue for the State of Alaska, which is the source of this water, but for Congress and their resistance to allowing changes or specific use exemptions in the old Jones Act – which we are working to amend. Population and growth in an area of little natural precipitation is the real problem, but without significant changes in the United States Congress and long existing water legal frameworks, or dramatic political change, let alone the host of theological and philosophical issues that become involved, we do not anticipate any real opposing success in this area. In the meantime, it's always important to remember water is needed not only for people but for economies to survive – around the world.

Finally, let us be honest. This is a new global generational venture and anything, with respect to challenges, is possible. The facts of the enormity of water in Southeastern Alaska and its millennium (for at least 30,000,000 years) of recharges of both land and sea and ice are well known. The bulk export of millions of gallons a day from one source will not in any measurable way harm anything other than the millions of human beings without adequate water to survive.

Most have no idea that the cry about Global Warming now called Climate Change is not scientifically valid in any way if you understand the known history of glaciers – especially in SE Alaska. We have more glaciers 'growing' then retreating. In addition, the snow fall on the back of glaciers is increasing mass more than replacing facial melt. But do not tell anyone. This is insider intel.

One of the more humorous charges is that the spillage from a tanker of freshwater into the open sea may harm sea life. This charge is on its face ignorant of the realities of the science of water and sea storms including massive amounts of rain fall. Freshwater briefly floats on the surface of oceans slowly dispersing down into the sea. A freshwater spill from a VMaxx tanker would not be measurable against the enormity of seasonal rain and snow.

No, we are confident that we, and the State of Alaska, who obviously have skin in this game, can withstand any environmental charges of "harm". The secret is the water source and the hydrologic region within which it is taken.

18 RISKS

18.1 Federalism

One of the most interesting notions voiced is the possible "federalization" of Alaska's waters

We heard that: *"Once people see use exporting American water to China, Obama will federalize it."* This was an actual statement from an interested, well educated, but politically uninformed, party. We pay attention to such, as these kinds of statements raise out of groups at international and national meetings.

Although some might think this a legitimate concern given the political attitudes of Congress and the former Obama Administration and new Bidon Administration, the realities are far from this becoming a reality. Mr. Davidge has decades of experience in Wash. DC and other political zones and keeps an ear to the DC games regularly.

First, **west** of the Mississippi, <u>all water</u> (surface, ground, snow, ice) unless <u>previously</u> withdrawn by the Federal government at the time of statehood, belongs to the States subject to its/state appropriation for "beneficial uses". That is the law affective at statehood for all states west of the Mississippi. In addition, "Western Water Law" is "first in time first in right" which clearly states that once appropriated by the state it is a <u>private "property right"</u> and is thus protected in American law including the Federal Constitution. Federal and State Case law also has over one hundred years of history of protecting water as a private property right.

Further, it is most unlikely the western states would allow any of their representatives in Washington D.C. to vote to sustain such a move. If you remember the "Sage Brush Rebellion" by 13 western states, this would be "The Water Revolution". State ownership and control of 'their' water is a big political deal in all states west of the Mississippi. It is not likely to ever change.

East of the Mississippi there are different laws, but even there it is not likely that state waters could be "federalized" without the consent of the state and then privately held water would have to be acquired – paid for- at fair market value. The federal government does not have enough money, even with its printer in Treasury.

18.2   Other Risks

Weather is always a risk issue that must me consider in delivery contracts. We must ensure that all delivery contracts have enough grace time in any schedule to allow for delays due to severe weather.

Terrorism is also a risk. The contamination of a conveyance technology, even tankers, can put that system out of business for some time. The security of these conveyances is an extremely high priority. Additionally, with the growing global value of water, these conveyances may be subject to pirates, especially in the Middle East. But we know a lot of retired Army Special Forces and Navy Seals who have chosen Alaska as their new home.

The ***failure to pay*** is an issue that must be considered in our business strategy. This potential is obviously one of the concerns in writing a contract with a prospective buyer and intuitively demands the closest attention to the capability and seriousness of the buyer. The problem is that once the water is delivered, it cannot be taken back if payment is not made. In the Delivery Contract, time of payment for water received is an essential issue that must be carefully addressed. You cannot expect the buyer to pay for water not yet received, nor water that has been received but not yet "cleared" based on a water quality test. But we certainly must insist in immediate payment for water "accepted". This is one of the reasons we prefer the buyer to provide their own conveyance, we get paid at the point of load before the tanker can depart. Additionally most pay in advance by purchasing a volume of water in 'futures'.

These payment problems are normally addressed through a down payment or 'futures investment' by the buyer that is held in escrow and drawn down upon market acceptance. This way we know the funds are in the bank before we release or deliver any water and automatically get paid once the water is accepted.

Again, the AQUEOUS preference is that all water is paid for at 'lift'. Once a tanker is filled and that shipment is certified, the payment should be made. NOT at point of delivery in markets. This is obviously a keen issue in our Delivery Contract negotiations, but those in market who want this water will concede as they have little option, and our performance should help them be comfortable with this payment process.

18.3 Legal Problems

The more probable challenge will be in the development and negotiation of purchase and delivery contracts in other parts of the world. **Having an international law firm with exceptional connections in key regional markets is essential.** The U.S. State and Commerce Departments and in some cases the CIA offer several ways to help in foreign markets including the location and reference of local competent attorneys. The development of such relationships is critical to our long-term business strategy. As a member of the World Trade Center and our use of Gold Passes via the State Department and our direct connection with the buyers in local markets, American Ambassadors have always been good for us and market partners.

This will not be an immediate issue, but one that will need to be addressed within the life of this plan. At least a DRAFT Delivery Contract needs to be developed that can work as a base from which to negotiate. Keep in mind that our goal is for water to become a global commodity of which we own significant sources. As a global commodity the sale/purchase of water futures is the tool best used. This effort is now underway by AQUEOUS International in the form of an outline of issues to be address/resolved. Bulk water is already being traded in open markets in California and other parts of the world. It will grow quickly once our tankers are in operation and the world discovers it economically viable.

18.4 Adverse Media

It is unlikely that the bulk export of natural water for bottling will generate serious interest by national or international environmental groups unless we tell them about it. Certainly, anything is possible, but we recommend we get contracts in hand, have our sources 'under paper' permit/rights, and move water – and we do all that without alerting all the wackos across the planet. Our marketing must be in market and retail level sales, not international news sources.

The more likely challenge will be against the plastic containers used by most bottlers in the sale and distribution of water. This is already under change within the global industry as many are now moving to use a highly biodegradable type of bottle made from a corn extract or canned or boxed water. We think boxed water at half gallon or greater is the most likely in the next two decades. Every other liquid is contained on the retail shelf, we are already in discussion of how to make boxed water retail attractive.

Thank You. If you are at this point in your review of this plan, you are one of the few. We understand that this is long and extremely detailed often repetitive plan but starting a new global renewable commodity we believe requires lots of detail and the beginning of trust between market partners, investors, and our company. If you have **any** questions or concerns with anything in this document, we ask you to call or email us immediately. We are confident we can answer you concerns. If not, we need to do some more work.

ABOUT THE AUTHOR

RIC DAVIDGE, MPA

1509 H Street, Unit A, Anchorage

Alaska 99501, International

PROFESSIONAL EXPERIENCE *(detail)*

2015 – Present **Author, *'The Second American Revolution – first 100 days'*** (Page Publishing) Still available on Amazon Books now rated 5 stars. Offers a very specific plan to actually achieve substantive structural government reform at the federal and state levels in 5 years back to and within the enumerated federal powers as were limited in Article 1 Section 8 of the Constitution, and again respecting the sovereignty of states (written in 2014) Fifteen new books in development: **'EQUAL?'** explores the historic effort by women to achieve legal, political, cultural, and personal equality – and what went wrong. **'Stopping America's slide into Socialism'** outlines, using Alaska as an example against the New Zealand miracle, how states & municipal governments are going bankrupt (12 states in America are totally bankrupt with little options in recovery, likely 3 more in 2023 and over 100 municipalities in just past decade) and then there is the federal debt and huge unfunded mandates that now exceeds the GDP of America. It is an old and well-known formula. Socialist over promise free stuff they have no way of paying for, then they blame the Republicans when the folks don't get 'em. I show how states can reduce cost, increase efficiencies, and really revolutionize how they provide services with responsible 'limited' competent governance. **Homelessness 2020,** now in its 3nd edition, 5th printing locally. This is a MPA formatted book: The Problem is, Context, Legal Constraints, Solutions. This book has gone national with Governors, Mayors, City Councils from across the nation asking for copies. It is printed in Anchorage. The new update has 5 new significant chapters. I also have **nine other new books on all things public policy** from the Alaskan portal. It is structured as a working library for <u>serious</u> conservatives especially candidates I have vetted – a tool with substantive research, critical context, and rational options. Latest release is **"What's wrong with Alaska's schools"** that has also gone national and even international. I'm delighted with my new publisher and am working hard on two new major books: **Reconstructing America's Republic of States**, and **The Future of Water in Pacific Markets.** The water book should do well, no one else is working on this that I can find.

9/94 – Present   **AQUEOUS, International, Inc.**   *Founder/Chairman/ President* (full time)

"***Alaska's Water Czar***" Anchorage Daily News, U.S. Water News, and Alaska's World Trade Center. Business Week calls him, **"One of America's new water cowboys."** Currently involved in the identification, analysis, and acquisition of 100 Billion gallons a year of high value pristine Alaskan <u>*natural glacial*</u> water for bulk export to world markets in desperate need of 'natural freshwater' for high quality beverages, bottled water, manufacturing, muni-mix, and science. This venture includes a relationship with an Alaska Native corporation to develop a new global multi-generational renewable industry in southeast Alaska that will not only provide jobs at every level but new science in the understanding of the differences in the very nature of waters of the world. AQUEOUS is also securing other high value natural freshwater sources in other areas of the world close to insistent markets.

A founding member of World Water, SA. AQUEOUS continues to provide consulting services to foreign and domestic companies developing bulk and bottled water exports in all areas of the world. Principle clients include companies in Saudi Arabia, Taiwan, Anchorage, Washington, Canada, South Korea, Tanzania, Japan, Mexico, California, Las Vegas, Philippines. Member, International Bottled Water Assoc., International Desalinization

Assoc., former member of the Western State's Water Council; past member, U.S. President's Commission on Western Water Policy. Retained by the International Joint Commission (US/Canada) on international water export issues. Articles in U.S. News and World Report, Newsweek Magazine, U.S. Water News, Business Week. Extensive TV, radio interviews in US, Canada, Europe. AQUEOUS International, Inc. is an active member of the World Trade Center, Alaska.

3/19 – Present **Alaska Roundtable** (a nonpartisan conservative caucus) *Founder/Referee*

Out of growing frustration with the Alaska Republican Party, with Republicans who are elected and then join the Democrats in governing and a party that refused to apply its rules to such RINOs, and a need to round up the highest qualified proven experts in public policy issues across Alaska, we founded this group. We are a closed room that encourages VERY honest frank discussions on all things Alaska. In less than a year we had a significant impact on state and municipal policies, public management, and elections. Why? Because we aggressively find, recruit, vet and then develop prospective candidates, and we are tough – not judgmental, just tough. We focus on 'development' so they have substance and know how to win especially in debates. We have candidates withdraw after our meetings deciding they need to learn more before running. We publish extensive booklets on Alaska policy, and we identify 'The Problems' (root causes) and offer viable solutions. These are tools for conservatives. Some of these booklets are on our website AlaskaRoundtable.com and will soon be published to be sold separately or as a library. We provide these to all candidates we decide to develop at no charge. Our book on Homelessness is now in its third edition and we get requests from all over the country for copies. Once they are candidates, we back away and let them win with their own campaign team, although most come to our weekly meetings or ZOOM - they say they learn more in our weekly 90 minutes then in any other forum.

2020- Present/Restarted, Alaska Oriental Gardens, which I originally began in 1995 as a Japanese garden designer, construction supervisor, and consultant. My garden on H Street is, according to the Japanese Ambassador who visited, the only authentic Japanese garden in Alaska. It has now been featured in national magazines and was for some time on the city garden tour, but their behavior in the garden was often not appropriate, and that relationship was terminated. Now I provide private tours, the Anchorage School for Japanese Emersion often visit, and various VIPs request tours especially visitors from Asia. The garden has unique trees and plants from Asia, but I try and use as much of Alaska as possible.

7/16-8/0/21 Midtown Community Council, Municipality of Anchorage
Elected President 11/20, formerly Vice President (volunteer/part time) Reelected VP in 2022 at request of Board. Provided an incredibly detailed controversial report on the so called 'homeless problem' in Anchorage. Why this happens and what the city and state need to do to fix it – but remain stuck on process. Well endorsed by several Community Councils and other groups. It restarted real discussion on what is NOT being done, like addressing root cause not just symptoms, and where all the money has gone. Initiated state legislative Anchorage delegation and Anchorage Assembly joint hearings to review progress et al. The problem is there is still no – no attention to connecting the homeless with jobs or education/skill development - the only proven cures for poverty. This booklet is being published by the AK Roundtable and has gained national attention. Had to resign because I moved my office to South Addition and no longer own any property in the Midtown district.

11/17-12/18 Dunleavy for Governor
External advisor to Senator Mike Dunleavy on general policy issues. After his election to Governor, helped as a member of the Budget Policy team during the Transition. Provide ongoing general policy and political advice when I see it needed and when asked.

2/18 – 12/20 National Museum on Arctic Warfare, Inc Corporate Secretary/Director (volunteer).

> Served as Interim President/CEO for about a year initially Founding Secretary/Director   This is a project underwritten by the AVF for about a year, until they got their our own C3. Building a new internationally significant museum and artifact curation facility telling the untold story of the invasion and yearlong occupation of parts of Alaska in WWII. All within our effort to also tell the boarder story of Alaska's past, present, and future significance in our national and global struggle for peace. This is not the same old museum idea, we fly, we teach, we rebuild returning veterans and get them employed in good technical jobs. Providing the work/talent for this adventure is the Veteran Intern Program (VIP) to help transitioning veterans and reduce veteran suicide. This museum is not just glass cases with old stuff in them. The best pay for us are blisters.

9/16 – 12/16 Trump for President Campaign Chairman, Alaska Veterans for Trump.

> Member of the Presidential Advisory Board for Trump until end of term.

9/08 – 2019   CY Investments, LLC Project Manager/CESCL/ Management Consultant (private

> contract/part time)   Overall project management responsibility for many residential land development projects in Alaska with development budgets in the multiple millions. Other clients include Driftwood Bay, LLC, Potter Creek Development, LLC, Huffman Timbers, LLC, Cedar Hills/Palmer. These projects represent well over 1,000 lots for workforce housing in neighborhoods in the Anchorage bowl and Eagle River areas.

3/07 – 1/08 ALASKA CULTURAL ARTIFACTS, LLC (partner died) Partner (part time) Why should the

> Alaska Native Artist who makes the art receive the smallest portion of the final sale price? We have crafted an alternative aggressive

Net Marketing platform that enables Alaskan craftsperson's to realize 80% of the final sale price of their work.

2/00 – 2/05 WORLD WATER, SA *President and Chairman* (full time) An international consortium,

Davidge provided leadership and vision in the development of this global company (Business, Strategic, and Financial Plans), registered in Luxembourg. The purpose of WW, SA was to acquire, market, and manage high quality bulk water sources to supply any area of the world. WW, SA is a consortium of NYK Line, Japan the largest shipping company in the world; ALJ Group, a multi-billion-dollar group located in Saudi Arabia with offices in most major cities in the world focusing on the development of new technologies and markets; NWS the largest bag/tug operating company in the world, AQUEOUS (formerly Alaska Water Exports), a holder of water rights; and other investors. This group is also linked to two other global companies responsible for the transportation of bulk water and research and development of the most cost-effective conveyance technology for the transoceanic shipment of bulk water.

6/00 – 6/10 NORTH STAR PRODUCTIONS, Inc. (501c3) *President, Producer/Director* (part time)

"Santa & Death," our best film. Only Alaskan Theatre Company that identifies, develops, produces, and promotes wholly Alaskan stage and feature film productions. Short Attention Span Theatre (SAST) is a well-known production of the company three times a year with 30+ new plays each season. The Alaska Playwright Workshop also helps recognize, develop, and promote new Alaskan playwrights. "The Magical History of Santa," created by Ric Davidge, is a Premier production that is destined to become a national Christmas classic. NSP develops new acting and technical theatrical and movie talent for the Alaskan market as well as New York and L. A. Co-producer of Premiere openings of original Alaska plays in Oklahoma City and Los Angeles. Teaches two classes a year, *"How to sell your Story to Hollywood"* and

"*How to Audition and Win.*" Annually produces, Alaska Veteran's Day "Heroes Concert", a nationally recognized event rebroadcast on statewide TV.

7/11 – 11/03 VIETNAM VETERANS OF AMREICA, GOVERNMENT AFFAIRS COMMITTEE,

National Chairman   (All federal legislation and interface with federal agencies)
   *Awarded VVA National Commendation Medal (8/13)

3/07 – 2020 ALASKA VETERANS POLITICAL ACTION COMMITTEE (VET PAC) Chairman

9/03 – Present ALASKA VETERANS FOUNDATION, Inc. (501 c 3) Founder and Chairman of the
   Board (part time) brochure available upon request

7/03 – 12/14 VETERAN'S AND MILITARY AFFAIRS COMMISSION, Municipality of Anchorage
   Chairman (2010-14) Vice Chairman (2005–06), Appointed: by Mayor Begich 7/2003, reappointed: Mayor Sullivan 2008

9/02 – 9/16 VIETNAM VETERANS OF AMERICA, Chapter 904 (501 c 19)
   President & Chairman, Director for Legislative Affairs - Member of the Board of Directors

6/07- 3/12 ALASKA VETERANS BUSINESS ALLIANCE (501c3) Founding Board Member/Chairman
   of Legislative Affairs Committee

6/06 – 8/10 VIETNAM VETERANS OF AMERICA, Alaska State Council (501 c 19)
   President/twice elected. 2013- present as Secretary. Chairman, Government Affairs 6/08 – present – Director of VVA Veteran Service Officers program w/state and national grants.

Ric Davidge, MPA

9/09 – 7/11   **Chairman, National Committee on Economic Opportunities, VVA National**

4/09 – 7/10   **Chairman, National Committee to End Veteran Homelessness, CSCP/VVA**

6/06 – 7/10   **NATIONAL CONFERENCE, STATE COUNCIL PRESIDENTS**, Vietnam Veterans of America

8/94 – 9/06   **ALASKA ORIENTAL GARDENS [exotic Asian landscapes/private & commercial properties]**
Owner (part-time, seasonal)
Landscape Artistry. I create unique natural environments. Design, installation. Ponds and water gardens, Japanese gardens, lighting, bonsai, and plant sculpturing. Not your usual Alaskan landscapes. Design award winner. H Street Garden, in downtown Anchorage, is a premier garden on the annual Alaska Botanical Garden tour and Anchorage Garden Tour. Gardens featured in national and international magazines such as Better Homes and Gardens/Country Gardens, spring 2005 with six-page article.

4/99 – 2/20   **CITY OF WASILLA   Sarah Palin, Mayor** *Economic Development* (full time staff/consultant)
Design and develop major elements of City long-range plans, develop programs, provide land use planning support to Planning Commission, City council and administration. Rewrite Comprehensive Plan (2000-2010). Identify grant sources, develop, solicit, and manage grant programs dealing with economic development planning. Develop and articulate city economic diversification strategic plans and provide annual progress reports. Develop and implement a Wasilla-specific database enabling quarterly reports for business and government and more creditable planning. Assist in resolving code enforcement conflicts. Responsible for significant public speaking and coalition building dealing with economic

development issues. Point person in the identification and solicitation of new business opportunities for the greater Wasilla area. Lead person in the development and planning of the city industrial park and airport.

6/96 – 4/99   **HIGHLAND HOMES**   *Owner/Residential General Contractor* (part-time)

Built luxury ($300,000 and up) homes in the Anchorage market. This is just something I had always wanted to do. My grandfather was a master builder in Hollywood, CA and I started working for him when I was small – collecting and straightening nails. He was a Scotsman and with the Black Watch in WWII – wounded twice, a sergeant. He was a great model for me.

11/89 – 7/09   **THE GREAT ALASKA-YUKON SHOWCASE**   *Director, Alaska Operations* (part-time/contract)

A private venture with associates in Alaska and Orlando, Florida. Responsible for all contacts in Alaska, sponsorship agreements, investor communications, interactions with public officials. The project provides a high-profile representation of Alaska and the Yukon for marketing year-round visitation, products, and training opportunities in the visitor industry. Contract for all talent and personalities for this venue awarded to Alaska North Star Productions, Inc.

2/95 - 3/89   **EASTER SEAL SOCIETY OF ALASKA**   *Director, Program Services* (volunteer)

Prepare and administer budget for program areas. Hire, train, supervise and evaluate staff in all program areas; develop new programs with focus on unmet needs; grant writing and management; coordinate planning process for short and long-term goals; evaluate supervised programs in relationship to community needs and relevant to Easter Seals' goals; develop contracts with other agencies to provide services unique to Easter Seals' expertise; act as a spokesperson for Easter Seals. Created and developed nationally unique program addressing needs of FAS/FAE at risk young adults in transition from home/school to jobs and independent living. Developed and received state certification for Care Coordinator Services and Respite Care programs. Expanded camping programs. Awarded Dole Foundation grant.

6/94 - 1/96   **FORTUNE PROPERTIES**   *Management Consultant* (part-time/contract)

Assist in development/implementation of corporate strategic plan to increase market share and diversify corporate economic base. Assist in development of media presentations and marketing programs. Develop/facilitate management training for professional staff. Review and develop cost containment and management systems. Assist in recruitment of top producers. Advise President/CEO on day-to-day management concerns and strategies. Broker all printing and manage signage contracts.

7/94 - 10/95 **CY INVESTMENTS, LLC** *Property Manager* (part-time) Responsible for full range of property management of all rental properties. Helping the wife.

7/94 - 9/94 **ALASKA DEPARTMENT OF NATURAL RESOURCES**
Assistant to the Commissioner

Led special initiative to resolve Title Navigability conflicts between the US Dept of Interior and the State of Alaska. Negotiate settlement agreement on approximately 10,000 streams and over 1 million lakes. Served as Department Ethics Officer. Review and recommend action by the Commissioner on all Contests and Appeals within the Department. Develop a work plan to implement the transfer of over half a million acres of state land to agricultural production consistent with the Mental Health Lands Trust decision. Review status of Title 38 revisions and prepare recommendations for new legislative package. Coordinate all ANILCA/CSU activities for the Department. Continue as principle contact on all water marketing for the State.

7/91 - 7/94 ALASKA DEPARTMENT OF NATURAL RESOURCES
Director of Water & Chief, Alaska

Hydrologic Survey   Appointed Alaska's first Director of Water by the Governor. Responsible for the development, articulation and presentation of Alaska water policy and the development, organization, and management of the new Division of Water with a budget of $5 million. The Alaska Division of Water managed an estimated 40% of our Nation's water resources including over 3 million lakes and an estimated 20,000 streams. These responsibilities included the State Water Policy and Management Strategy; issuing water rights; administering the dam safety program; rendering title navigability determinations and Quiet Title actions in federal court asserting ownership of 14 million acres of submerged lands; surveying, collecting and distributing water resource data related to the quantity and quality of surface, ground and coastal waters of Alaska; coordinating water related data collection and management activities with other agencies; providing support to the State Water Board; and advocating responsible water development including water exports. Represented the Governor at the Western States Water Council (17 western states). Served as Co-chair of the Alaska Water Management Council representing 22 state/federal agencies and University of Alaska. Key advocate for creation of this innovative council. Appointed to the President's Commission on Western Water Policy.

The Division of Water had four sections: The Alaska Hydrologic Survey including the State Water Lab, Water Management and Development, Title Navigability, and Dam Safety with offices in Anchorage, Fairbanks, Juneau, and Wasilla. Additional responsibilities assigned included coordinating all DNR interactions with federal agencies, development of DNR's first Strategic Plan and Tactical Plans, and first DNR annual report of accomplishments under the strategic plan.

The Division increased its off-budget revenue by over 300%

in 18 months. During the 1994 Legislative Session significant budget cuts were initiated across state agencies. To reduce administrative costs within the Department of Natural Resources several Divisions were consolidated. This included the Division of Water with the Division of Mining.

2/91 - 7/91 OFFICE OF THE GOVERNOR *Assistant to the Governor* for Policy and Legislation

Responsible for the development, articulation, and presentation of State policies in a wide range of spectrums. Principle staff assistant to the Governor's Advisory Council on Subsistence. Governor's liaison with the State House Republican Minority. Responsible for review and analysis of legislation and proposed policies. Responsible for a range of areas dealing with State and Federal relations.

3/89 - 2/91 ENVIRONMENTAL SERVICES, LTD. *Director*, Planning, Permitting and Government Affairs

Executive position responsible for all planning, permitting and government affairs for the clients of ESL. Clients ranged from small mining to world class mining operations, underground storage tank management for international firms in Alaska, Seattle and Canada, oil support industries, major oil companies, commercial fishing, commercial real estate development in urban and rural areas, oil spill clean up (Exxon Valdez to small scale). Development and articulation of client corporate policies on cultural and natural resources. Monitor and actively participate in all phases of state and local government regulatory or statutory activities that may impact clients. Negotiated acquisition agreements between state and private mineral right owners. Initiated Major Joint Venture to provide environmental consulting services in areas of the former Soviet Union. Exxon/Valdez Oil Spill - Responsible for environmental compliance for VECO (ESL Client) in the Gulf of Alaska. Designed and supervised second largest most diversified waste receiving, sorting, packaging, and transporting system for the spill.

1989 - 2/91 ALASKA ECONOMIC DEVELOPMENT CORPORATION
President (part-time)

>Participating in the financing and development of coal leases in the Beluga area. This was a limited partnership involved with European financing and an investment group in Alaska. AIDC was also involved in economic diversification for the opportunity in Beluga including wood fibers, agricultural products, limestone, gravel exports, etc... The coal project had State and Federal support.

6/86 - 2/91 ALASKA PUBLIC POLICY CONSULTING *President* (part-time)

>Clients included; American Land Alliance, the National Inholders Association, Domestic Policy Council in the White House on national energy policy and national security. Local contracts involved policy development and articulation for legislators, legislative committees, statewide interest groups, and nonprofit organizations.

1988 - 11/90 ALASKA PROFESSIONAL SPORTSMEN'S ASSOCIATION *Executive Director* (part-time)

>Developed an aggressive professional membership association in partnership with several outdoor recreation groups and companies. Solicited and received corporate sponsors and national product endorsements for conservation. Developed, articulated, and presented legislation and testimony on a wide variety of fish and game and outdoor recreation issues to state and federal legislatures and agencies. Developed and published Code of Ethics. Subsistence, fish, and game management, permitting systems, agency personnel, conflict resolution for members with government agency problems. Annual conferences, designed, wrote and published newsletter. Membership development and financial management.

5/88 - 3/89 SUSITNA INDUSTRIES, Inc. *Vice President, Assistant General Manager (full-time)*

Executive position responsible for the exploration of new development opportunities in port, industrial park, natural resource exports and end product manufacturing (coal, wood fibers, agricultural products, limestone). Planned, organized, and directed the search for new opportunities. Coordinated government relations efforts, prepared analytical papers for CEO and Board of Directors, coordinated work by contract consultants in a wide range of fields. Prepared and presented development proposals ($250 million to $1.5 billion range). Negotiated land leases, purchases, and sublets with private owners and government bodies. Assisted in the identification and development of international financing of projects. Responsible for corporate long range planning. Significant public speaking responsibilities in the presentation of projects. Responsible for community relations.

5/87 - 4/88 MATANUSKA SUSITNA BOROUGH, DEVELOPMENT SERVICES DEPARTMENT

Director (full-time) Executive position responsible for six divisions with 30 employees. Planned, organized and directed the Platting, Public Lands, Planning, Parks and Recreation, Cultural Resources and Code Compliance Division. Administered contracts, requests for proposals and represented the Borough in negotiations and economic development. Significant public speaking responsibilities. Coordinated the Borough's economic development and long-range planning, capital projects, budget development and legislative interactions. Coordinated activities of citizen advisory groups and was ex-officio member of Planning Commission. Set and met departmental goals and objectives. Acting Borough Manager in the absence of the Manager. Member of the Interdisciplinary Planning Team of Susitna Forest Management Plan. Developed and headed successful International Trade Missions to Finland and South Korea.

1/87 - 6/87 CITIZEN'S COALITION FOR TORT REFORM *Executive Director* (full-time)

Responsible for development of statewide Coalition. Created, prepared, and published all publications, including monthly newsletters and weekly Legislative updates. Organized initiatives by Coalition to impact political decisions in Juneau and Washington, D.C., on significant reform of tort laws within the state and nation. Organized and facilitated visits by national figures throughout Alaska on behalf of Tort Reform. Prepared and implemented quarter-million dollar budget, developed, and directed fund raising, organizational development, base broadening, and representation responsibilities to State Legislature, public interest groups and national organizations. Got them out of debt in 6 months.

1/86 - 6/86 U.S. FISH AND WILDLIFE SERVICE (FWS) *Assistant to the Director (full-time)*

Assigned leadership responsibilities for extremely sensitive issues in Alaska. Chairman, Staff Work Group responsible for the development and implementation of a Federal Subsistence Resource Management Program for Alaska; Chairman, RS2477 Task Force including federal, states, native corporations, and state legislators.

3/81 - 12/85 ASSISTANT SECRETARY OF UNITED STATES, DEPARTMENT OF INTERIOR

Special Assistant (full-time) By request of the President, appointed Special Assistant to the Assistant Secretary of Interior for Fish, Wildlife and Parks. Responsible for all National Park Service programs; developed Management by Objectives program for DOI for Secretary, developed and implemented President's national policies on land protection, historic preservation, and recreation. **Chairman of Federal Land Policy Group** (DOI & DOA), **Federal Commissioner** to the New Jersey Pinelands Commission, **Chairman of the Department's National Task Force on Designation of Undeveloped Coastal Barriers**. Secretary of the Interior's representative on the **President's Council for Historic Preservation** and **Board of Directors, National Trust for Historic Preservation**. Represented the Secretary and Administration at National Conferences addressing park, wildlife, historic preservation, and recreation policies. Received awards for performance from Secretary and special recognition by U.S. Congress for leadership in the Coastal Barriers program and in development of alternative land protection programs. Often served as **Acting Assistant Secretary**. Requested transfer back to Alaska, June 1983. Responsibilities included work with the Alaska Land Use Council, Governor and four federal land agency directors. Principally responsible for land use planning, issue coordination, policy review and liaison between the Fish and Wildlife Service, the National Park Service, local governments, and the State of Alaska. Wrote planning guidelines for all ANILCA related planning of NPS and FWS units (over 100 plans). Responsible for studies and investigations of organizational and personnel issues for the Assistant Secretary of Interior. Worked closely with the Office of the Governor, the Citizens Advisory Commission on Federal Areas, the Commissioners of DNR, DEC, ADF&G and the Office of Management and Budget.

3/80 - 2/81 **INDEPENDENT CONSULTANT, WASH, D.C.** (full-time)
Director of Governmental Affairs for National Association of Property Owners, Managing Director for National Inholders Association, and Program Director for Institute for Human Rights Research. Private consultant to several small property and homeowner groups in many states. Negotiated several pieces of federal legislation to compensate federal acquisition of private lands. Consultant on land issues to the Republican Presidential Congressional Caucus, member of the Conservation and Sportsman Committee to advise President Reagan during the 1980 campaign. Outside consultant to the President's Transition Team for the Interior Department and the White House on national land use policies. Consultant/leader in the Sagebrush Rebellion.

8/78 - 3/80 SENATOR TED STEVENS, WASHINGTON, D.C. *Staff Assistant (full-time)*

Responsible for special projects including a major national investigation of federal land acquisition policies and practices focusing on the Departments of Interior and Agriculture. Responsible for CETA, Job Corps, CEDC, all legislation creating new or expanding federal areas outside Alaska, Tourism, Department of Justice Lands Section. Routinely reviewed legislation, prepared public policy papers, developed issue briefs, prepared floor statements and speeches, made voting recommendations on national legislation to the Senator. A transformative experience working with and learning from an Alaskan icon.

1974 – 1978 Leadership/Statewide Education Organizations and UNIVERSITY OF ALASKA (UA)

Chief Executive, facilitator, and lobbyist for Alaska Student Lobby, representing over 40,000 students. Member of UA Board of Regents by Gubernatorial appointment. Chairman UA Statewide Budget Review Committee. President of UA Fairbanks Associated Students. Recognized statewide leader in Alaska Post Secondary Education. Faculty with Tanana Valley Community College, and Juneau/Douglas Community College. I was a full time student who also taught dance, choreographed all stage productions, and was on full scholarship for all undergrad and postgrad work. I worked the entire time I was a full time student.

1975 – 1976 FAIRBANKS NORTH STAR BOROUGH Chairman, Public Transportation Commission

Appointed by the Mayor as a member of the borough's first public transportation commission and then later appointed as Chairman. Developed the first borough wide public transportation system to be phased in over a decade allowing residents to use buses for public transportation. Downtown city bus was offered free to passengers for the first year. This program was recognized by a number of national transportation organizations for its innovations.

1975 – 1976 PIPELINE IMPACT INFORMATION OFFICE *Government Intern* (part-time)

Researched and wrote on socio-economic impacts of Alaska pipeline construction on the Fairbanks North Star Borough.

COMMUNITY SERVICE

- Municipality of Anchorage, Mayor's BEAR Workgroup (keeping our troops in AK & expanding Anchorage's and our Arctic's military defense missions)
- Municipality of Anchorage, Mayor's Transitions Teams (Begich, Sullivan, Berkowitz)
- Chairman, District 20 – Alaska Republican Party (2010 – 2020)
- Member, AKRP State Central Committee (2012 – 2020)
- Chairman, Alaska Republican Party Platform Committee (2016-2020)
- Co-Chairman, AK Republican Party Committee on National Defense, and Veterans Affairs (2012- 2020)
- Delegate, RNC National Convention, Member Platform Committee/2012
- Campaign Chairman AK, Gingrich for President Campaign 2012
- Anchorage/Seattle Economic Development Council
- Alaska Veterans Foundation, Inc. (Founder/Chairman 2003 – 2021)
- Veterans and Military Affairs Commission, Municipality of Anchorage (too long)
- Vietnam Veterans of American, Chapter 904
- VetCity, Inc. [ending veteran homelessness in Alaska]
- Vietnam Veterans of America, Alaska State Council
- Commonwealth North Committee on State/Federal Relations
- Commonwealth North Committee on ANWR
- Commonwealth North Committee on Urban/Rural Issues
- Board of Directors, Resource Development Council
- Board of Directors, Families First Partnership

- Anchorage Economic Development Corporation
- Founding Member, Republican Presidential Task Force (Honor Roll)

2016 Municipality of Anchorage, Bid Review Board (appointed by Mayor)

2015 Mayor's BEAR (Base Economic Assessment Recovery) team in response to proposed troop reductions at Ft Richardson – we stopped the loss of troops in Anchorage

2015 Mayor's Transition Team (Berkowitz/Community Development)

2013 Awarded - VVA National Commendation Medal

2009 Mayor's Transition Team (Sullivan/Military and Veterans Issues)

2006, 2005 Twice Awarded - Congressional Order of Merit (NRCC)

2004 Awarded - Ronald Reagan Republican Gold Medal (NRCC)

2004 – 2014 Member (Chairman 2010, former Vice Chairman), Municipality of Anchorage, Commission on Military and Veterans Affairs

2006 – 2008 Chairman, Committee for the design, funding, construction, and dedication of the new Municipal Veterans Memorial in the Delany Park Strip, Anchorage

2003 – 2010 Producer (writer & director) Alaska Veterans Day Concert, Alaska Center for the Performing Arts

2003 Mayor's Transition Team (Begich/Military and Veterans Issues)

2002 – present Board of Directors, Chairman/CEO Alaska Veterans Foundation, Inc. (501 c 3)

7/02 – 2018 Board of Directors, President, former 1st Vice President, Chairman of By Laws Committee, and Chairman Governmental and Legislative Affairs, Vietnam Veterans of America, Chapter 904

3/99 – 2001 Board of Directors, Legislative Committee Chairman, Families First Partnership, Inc.

8/99 – 11/99 *Chairman*, Trade and Development Subcommittee Alaska Commission on Privatization and Delivery of Government Services

Ric Davidge, MPA

6/93 - 11/95 *Member*, Board of Directors and Chairman, Legislative Committee, Friends of Children

9/90 - 11/90 *Statewide Issues Coordinator*, Walter Hickel for Governor Campaign

12/89 - 11/90 *Statewide Issues Coordinator*, Coghill for Lt. Governor Campaign

3/89 - 11/90 *Founding member*, ANWR Committee, Alaska Support Industry Alliance

6/86 - 5/87 *Founding member* Coastal Plain Committee, The Alaskan Coalition for American Energy Security

6/86 - 5/87 *Director*, Division of Lands, Resource Development Council

Responsible for development and articulation of State/Federal land use issues. Addressed such problems as ANWR Oil & Gas Exploration and development, and Federal/State area management plans, Chairman, ANWR Task Force. Alaska *Representative* of the American Land Alliance

6/86 - 8/86 *Issues Coordinator*, Walter Hickel for Governor Campaign, responsible for Development & Presentation of <u>all</u> issue papers and special interest group questionnaires. Prepared speeches for the candidate and functioned as a surrogate speaker.

1975 - 1976 *Chairman*, Commission on Public Transit, Fairbanks, Alaska

Developed and implemented new comprehensive public transportation system for the Fairbanks North Star Borough.

FORMER BOARDS

- Board of Regents, University of Alaska (statewide)
- Board of Directors, Resource Development Council

- Board of Directors, Arts Alaska
- Board of Directors, University of Alaska Alumni Association
- Board of Directors, Alaska Dance Theater
- Board of Directors, Asia International Corp.
- Board of Directors, Anchorage Community Theatre
- Board of Directors, Vietnam Veterans of America, 904
- Board of Directors. Vietnam Veterans of America, Alaska State Council
- Board of Directors, Alaska Veterans Foundation, Inc.
- Board of Directors, Alaska Military Heritage Museum

EDUCATION

- Pursuing **PhD** in government management (underway)
- Post Graduate work in Arctic Architecture and Engineering, Univ. of Alaska, Anch 1994
- **Master's in Public Administration (MPA)**, University of Alaska, Juneau, 1978
- Post Graduate work in Economics, Political Science, and Management, 1976
- **Bachelor of Arts (Music/Theatre)**, University of Alaska, Fairbanks, 1975

PROFESSIONAL ASSOCIATIONS

- Member, International Bottled Water Association (inactive)
- Member, International Desalinization Association (inactive)
- Member, Project Management Institute (inactive)
- Member, ICMA (International City Management Association) (inactive)
- Member, World Trade Center Alaska (inactive)

PERSONAL INFORMATION

Born March 20, 1946 (Southern California); Married; 3 grown children; 6 grandchildren Former professional singer, composer, producer, dancer, director, and choreographer. Dance Theater instructor and performance choreographer, Univ. of Alaska, Fairbanks. Three years under contract with the Alaska Light Opera Theater as Company Choreographer.

Worked with Andy Williams, Bob Hope, Red Skelton, Johnny Mann, Ann-Margaret, Juliet Prowse and others in stage, TV, and world tours. National radio talk show personality.

(Separate Theatre resume available only upon request – it confuses people)

Foreign Residencies: Vietnam, South Korea, Japan, and Hong Kong. I have traveled extensively in Asia and the Pacific Rim as well as Europe, the Middle East, northern Africa, New Zealand, Australia, Philippines, Guam, Albania, Cypress, etc .

Military Service: US Army 1963 to 1969   Viet Nam Combat Veteran: Medic, Flight Service/ATC, Admin

Honorable Discharge   Assigned: Fort Ord/Calif.; Fort Sam Houston/Texas; Fort Benning/Georgia; Fort Rucker/Alabama; Vietnam/An Khe, Plieku, Bong Song; 8th Army UN Command Headquarters, Seoul, Korea